BEYOND THE AGE OF OIL

BEYOND THE AGE OF OIL

The Myths, Realities, and Future of Fossil Fuels and Their Alternatives

LEONARDO MAUGERI

Translated from the Italian by Jonathan T. Hine Jr.

AN IMPRINT OF ABC-CLIO, LLC
Santa Barbara, California • Denver, Colorado • Oxford, England

Originally published in Italy in 2008 by Sperling & Kupfer Editori S.p.A. as *Con tutta l'energia possibile.*

First English-language edition published in the United States in 2010 by ABC-CLIO with revisions by author.

Library of Congress Cataloging-in-Publication Data

Maugeri, Leonardo, 1964–
Beyond the age of oil : the myths, realities, and future of fossil fuels and their alternatives / Leonardo Maugeri; translated from the Italian by Jonathan T. Hine Jr.
p. cm.
Includes bibliographical references and index.
ISBN 978-0-313-38171-3 (hard copy : alk. paper)—ISBN 978-0-313-38172-0 (ebook) 1. Power resources. 2. Renewable energy sources. 3. Petroleum reserves—Forecasting. 4. Energy industries. 5. Energy policy. I. Title.
TJ163.2.M37513 2010
333.79—dc22 2009048615

ISBN: 978-0-313-38171-3
EISBN: 978-0-313-38172-0

14 13 12 11 10 1 2 3 4 5

This book is also available on the World Wide Web as an eBook. Visit www.abc-clio.com for details.

Praeger
An Imprint of ABC-CLIO, LLC

ABC-CLIO, LLC
130 Cremona Drive, P.O. Box 1911
Santa Barbara, California 93116-1911

This book is printed on acid-free paper ∞

Manufactured in the United States of America

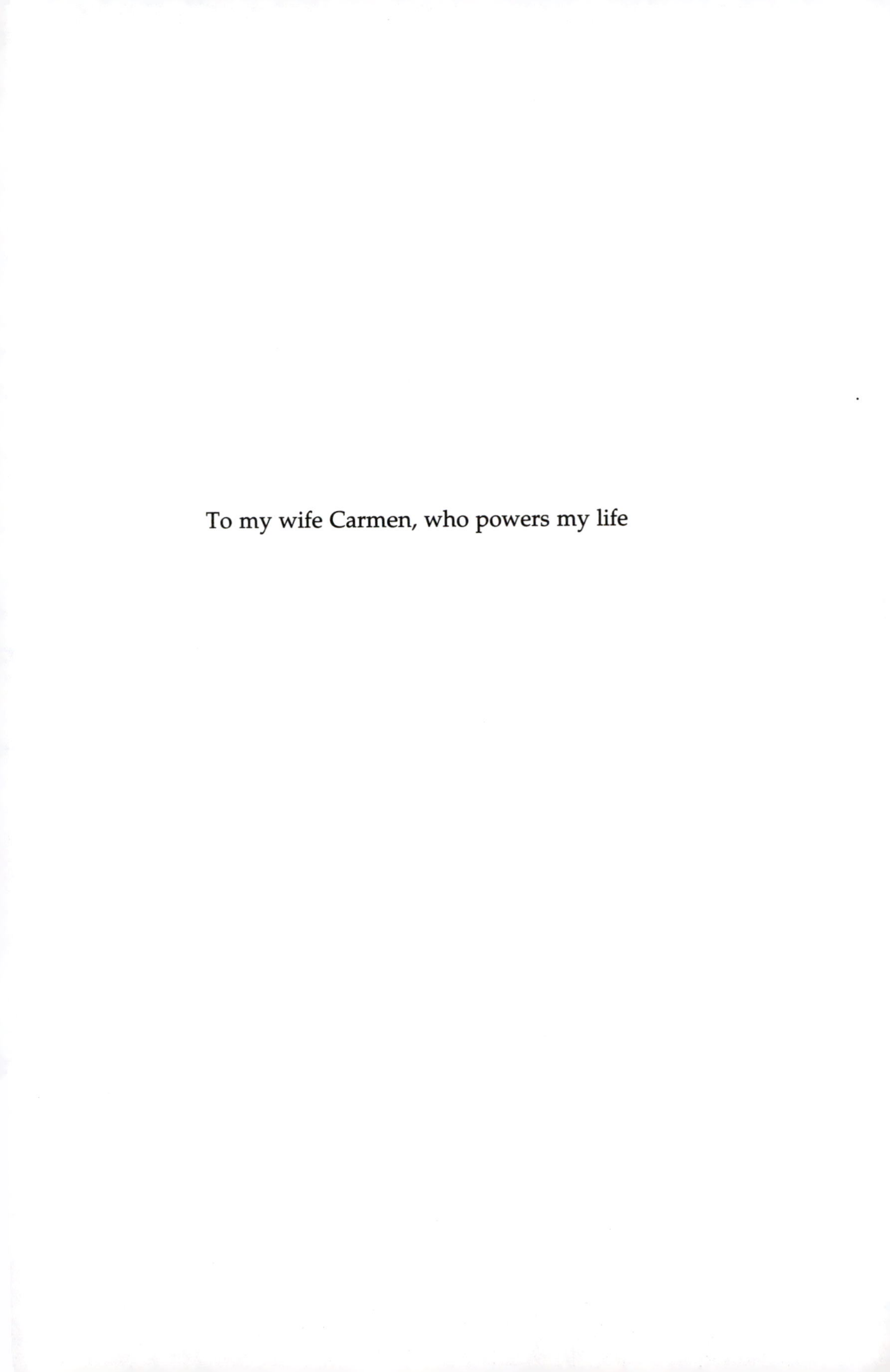

To my wife Carmen, who powers my life

Contents

Preface

This book is an enlarged and updated edition of a book that I began writing in Italian in 2007. It meets the same needs that drove me to write a book about all energy sources. Something was missing. For some time at work, I had been analyzing the potential and limitations of each source of energy. There was talk everywhere about energy, about the need to make an energy transition, about renewable sources, and about alternatives to fossil fuel (oil, coal, and gas). As I was preparing a presentation for a meeting at an American university on the real prospects of having all the energy we need, I noticed that the literature on the subject was not accessible to someone trying to develop an informed opinion. True, there were excellent textbooks on energy, written by excellent academics. Most of these spoke only to the initiated, those who already had a deep knowledge not only about energy but also about physics, chemistry, engineering, and economics.

On the other hand, dozens of books were coming out prepared by careless authors who confused installed power capacity with electrical generation. They underestimated or ignored the basic problems of energy and power density. They enthusiastically showed the high percentage increase in production from renewable sources of energy, without explaining that their starting points were so low that even doubling or tripling their production did not affect at all the energy scenario. They reported about fantastic inventions without having any cognition of technology. They even confused primary energy sources with secondary sources, such as hydrogen, which needs to be derived from a primary energy source. And this is only a small list from many other more or less egregious errors.

A lack of knowledge did not keep them from arrogantly pronouncing apocalyptic scenarios for fossil fuels (for petroleum in particular) or forecasting a wonderful future near at hand for renewable energy. Meanwhile, the most advanced laboratories in the world had been trying to bring about that renewable future for decades, without satisfactory results so far.

Then there were publications by competent but partisan scholars who, in their dedication to a single cause, had compromised the rigor of their analysis and the objectivity of their conclusions. With the exception of the academics, few had tried to provide a precise picture of the different sources of energy, concentrating rather on a single theme such as hydrogen or biofuels, or a category such as renewable resources. None offered a total vision of the energy possibilities, ranging from what we have now to what we could have in the future.

I have tried to fill this gap by writing a book that I hope will be easy to read. At the same time, I do not want to be accused of excessive simplification or, worse yet, intentional manipulation. I am aware that my work could be considered tainted because I happen to be a senior executive in an oil company (Eni SpA) and I have already written enough about petroleum, though usually from a contrarian point of view. Let me emphasize that I have no intention at all of defending petroleum.

By background and education, I am a rational environmentalist, irrepressibly curious and open to all possible alternatives to our present energy situation. Among my responsibilities at the company where I work is scientific and technological research and innovation, and one of my tasks is to guide the research and development of alternatives to petroleum. Therefore, I have dived enthusiastically into the task of devising a medium- to long-range program to do this. It is fascinating and stimulating work, although there is no lack of obstacles.

This work has introduced me to the laboratories of the most important universities and research centers in the world. It has given me a feel for our real prospects and the limitations that still confine us to the energy sources that are available today. It has led me to draw some conclusions, and to nurture many hopes for the future. True, the solution to our energy trap—i.e., our dependence on fossil fuels—is not just around the corner. But it could come if we devote ourselves to research and technological innovations as never before, with patience, seriousness, and cross-fertilization among different disciplines.

To explain all this is not simple. Energy is a very complicated issue, which makes it easy both for informed people to remain obscure in their narrative and for uninformed people to oversimplify any issue. I have tried my best to resolve these problems. Although I have tried to simplify the numbers and the technical explanations, numbers are essential to an objective discussion of the subject. For each source of energy, I have tried first to explain its nature, describe how we exploit it today, and summarize its history to help us understand the present situation. Then I have shown its potentials and limits, its advantages and disadvantages. Finally, I have tried to look at its possibilities in a not-too-distant future, about the year 2030.

The only exception to this approach is the chapter on petroleum, which opens the book. It is longer and structured differently than the other chapters, simply because in the modern history of energy, petroleum prices have always influenced trends for all other sources of energy. In particular, low oil prices have thwarted the expansion of renewable sources, energy efficiency, and scientific research into alternative energy sources. The dominant role of crude oil has made it the undisputed arbiter of any energy future. For this reason, I think it is essential to provide some fundamental guidelines to answer the many "why" questions. For example, why do petroleum prices seem to defy prediction, alternating between boom and bust cycles? Why is the fear of running out of petroleum such a recurrent (albeit unfounded) theme? Why do we still not know how much oil there really is?

To keep the numbers from bogging down the text, I have collected the principal data concerning each resource into tables at the end of that resource's chapter. To provide the most up-to-date information possible, I turned to reliable sources. However, those sources differ, because still today we lack a coherent, up-to-date, and dependable source that can give us a vision of all the sources of energy together. Thus, there may be slight mismatches between the numbers, but these small inconsistencies have no impact when comparing general trends and orders of magnitude.

Units of measure may cause problems. Frankly, they cause problems for the experts, too. To handle the different energy sources consistently, we need complex conversions into units that usually are not those used in talking about a particular energy source. I have tried to avoid these complications, and I hope that I have succeeded.

When I started writing this book, I planned to limit each source of energy to twenty pages for the most important sources and ten pages for the others. An issue as complex as energy could easily sap the ability of readers to reach the end of the book. With the help of my translator, Jonathan Hine, I have tried hard to summarize, while providing the most extensive explanation that I could. I added new parts and included the very latest data and significant events. Though I could not always fully keep my page-count promise to myself, I hope that I have covered each energy source in a bearable number of pages. I also hope that I have served you better with this effort.

Acknowledgments

As one writes a book, a debt of gratitude accumulates to all those who have helped to define the essentials, revise the draft, correct the proofs, do the research, and check the numbers. It is always difficult to keep track of the individuals who contributed, whether by a quick exchange of opinions or a well-developed contribution over time. To draft and complete this work, I counted on the advice, stimulus, criticism, and careful revision of many people, most of them my colleagues and coworkers.

Here are those who had the most direct impact on the individual chapters and parts of my work, in alphabetical order: Rita Calento, Marco Cecchini, Massimo Chindemi, Gianluca Chiodini, Elvira Di Sibio, Lorenzo Esposito Caserta, Tolga Hunturk, Giampiero Marcello, Salvatore Meli, Riccardo Mercuri, Franco Palermo, Cristiano Pattumelli, Carlo Perego, Manuela Rondoni, Marianna Russo, and Stefania Santomauro. To these I must of course add all those who work with me at Eni in the Department of Strategies and Development.

Above all, I owe special thanks to a small group of my collaborators who carefully checked the text several times using their specific expertise. They are Francesca Ferrazza, Valentina Garruto, Sabina Manca, Giuseppe Sammarco, Luigi Sampaolo, Claudia Squeglia, and Dario Speranza.

I have an abiding debt to various professors (and friends) of the Massachusetts Institute of Technology (MIT), who agreed to review specific chapters of the book notwithstanding their very important and challenging work of finding new solutions for our energy future. Sometimes we shared the same views; sometimes we differed. Their work was immensely precious to me, even when they criticized some of my positions. If I failed to capture the depth of their observations or to correct some point, I alone am to blame. I remain responsible in every way for the ideas and notions expressed in the various chapters that they reviewed. In alphabetical order, they are (with the chapter they read): Robert C. Armstrong (natural gas); John M. Deutch, (coal, carbon dioxide); John B. Heywood (biofuels); Robert van der Hilst (oil); Ernst J. Monitz (nuclear energy); and Jefferson W. Tester (geothermal energy). I am also grateful to Nicola De Blasio for maintaining the link with these professors and interacting with them.

I have an additional debt of gratitude for the help and advice I received from my friend Branko Terzic, former U.S. state and federal energy regulator who is now global regulatory policy leader for energy and resources at Deloitte Services. He kindly agreed to review the chapter on energy efficiency at the very last minute, and he did it graciously, meeting our tight deadline. I also must thank my brother Alessandro, who acted as an incredible fact-finder, indefatigably navigating among hundreds of sources. This new edition of a book that I originally wrote in Italian would have never been so fluent, precise, and to the point without the brilliant work of my translator, Jonathan T. Hine Jr. Working with a translator was a new experience for me. I wrote my previous book (*The Age of Oil*) in English, and I thought it was much simpler to think and write in English than to have everything translated and have to review the translation. But Jonathan went far beyond my expectations. We exchanged ideas constantly, and he was always ready to catch the essence of my original Italian text, as well as that of the many additions and corrections that transformed it into a new work.

Special recognition goes to two people who for years have helped me manage my office and my activities: Anna Laura De Francisci and Nadia Sturmann.

As always, I could never have begun or finished this work without the sweet, ever-present, and silent support of my wife Carmen. For the time that it took me away from her, she should be considered a co-author.

For all the help that others gave me, they are certainly not party to any mistakes. The opinions and judgments expressed in this book, in whatever form, are mine alone.

Abbreviations

AC	alternating current
BTU	British Thermal Unit
CAFE	corporate average fleet economy
CCGT	combined-cycle gas turbine
CCS	carbon capture and storage; carbon capture and sequestration
CERA	Cambridge Energy Research Associates
CIA	Central Intelligence Agency
CNG	compressed natural gas
CO_2	carbon dioxide
DC	direct current
DOE	Department of Energy (U.S.)
EGS	enhanced geothermal systems
EISA	Energy Independence and Security Act
Eni	formerly the acronym for *Ente Nazionale Idrocarburi* (a government agency), now the name of a privatized energy company
EOR	enhanced oil recovery
EPR	European Pressurized Water Reactor
EU	European Union
FAO	Food and Agricultural Organization (United Nations)
FERC	Federal Energy Regulatory Commission
FPC	Federal Power Commission (replaced by the FERC)
GECF	Gas Exporting Countries Forum

GDP	gross domestic product
GTL	gas-to-liquids (natural gas conversion technology)
GW	gigawatt(s)
HDR	hot dry rocks
IAEA	International Atomic Energy Agency
ICE	International Commodity Exchange
IEA	International Energy Agency
IGCC	integrated gasification combined cycle
IOR	improved oil recovery
IPCC	Intergovernmental Panel on Climate Change
JODI	Joint Oil Data Initiative
kcal	kilocalorie(s)
kWh	kilowatt-hour(s)
LED	light-emitting diode
LNG	liquefied natural gas
LPG	liquefied petroleum gas
MBD	million barrels per day
MBTU	Million British thermal units
MIT	Massachusetts Institute of Technology
MW	megawatt(s)
NASA	National Aeronautics and Space Administration
NYMEX	New York Mercantile Exchange
OAPEC	Organization of Arab Petroleum Exporting Countries
OECD	Organization for Economic Cooperation and Development
OPEC	Organization of Petroleum Exporting Countries
OTC	over-the-counter
SUV	sport utility vehicle
syngas	synthetic gas (product of the gasification of coal)
TEP	ton-equivalents of petroleum
TWh	terawatt-hour(s)
USGS	U.S. Geological Survey
WEC	World Energy Council
WTI	West Texas Intermediate

Introduction: All the Energy in the World

Without *fossil fuels* (petroleum, coal, and natural gas), the world would not be the way we know it today. Twenty-first-century humans could not live without these three nonrenewable sources. Today, petroleum, coal, and natural gas provide 80 percent of the world consumption of *primary energy* (energy found in natural resources before any human conversion or transformation). Fossil fuels yield energy directly through combustion. The dominance of fossil fuels is causing keen anxiety, and for good reason. The growing use of fossil fuels has a negative effect on the environment in which we live and on the climate of our planet. When they are burned, coal, petroleum, and natural gas (in that order of environmental impact) emit many pollutants and contribute to global warming. The statistics and forecasts lend themselves to heated debates among scientists. Some warn of a dark, catastrophic future, while others deny any danger.

That our planet is suffering from the effects of our species' dependence on fossil fuels is unquestionable. The most striking evidence that we have abused our planet is the amount of accumulated carbon dioxide (CO_2) in the atmosphere. From ice samples taken at different depths, we know that for about ten thousand years until the mid-eighteenth century, the amount of atmospheric carbon dioxide remained fairly stable at about 280 parts per million. In the last 250 years, however, the concentration jumped to about 390 parts per million, and it is still increasing. There are many other

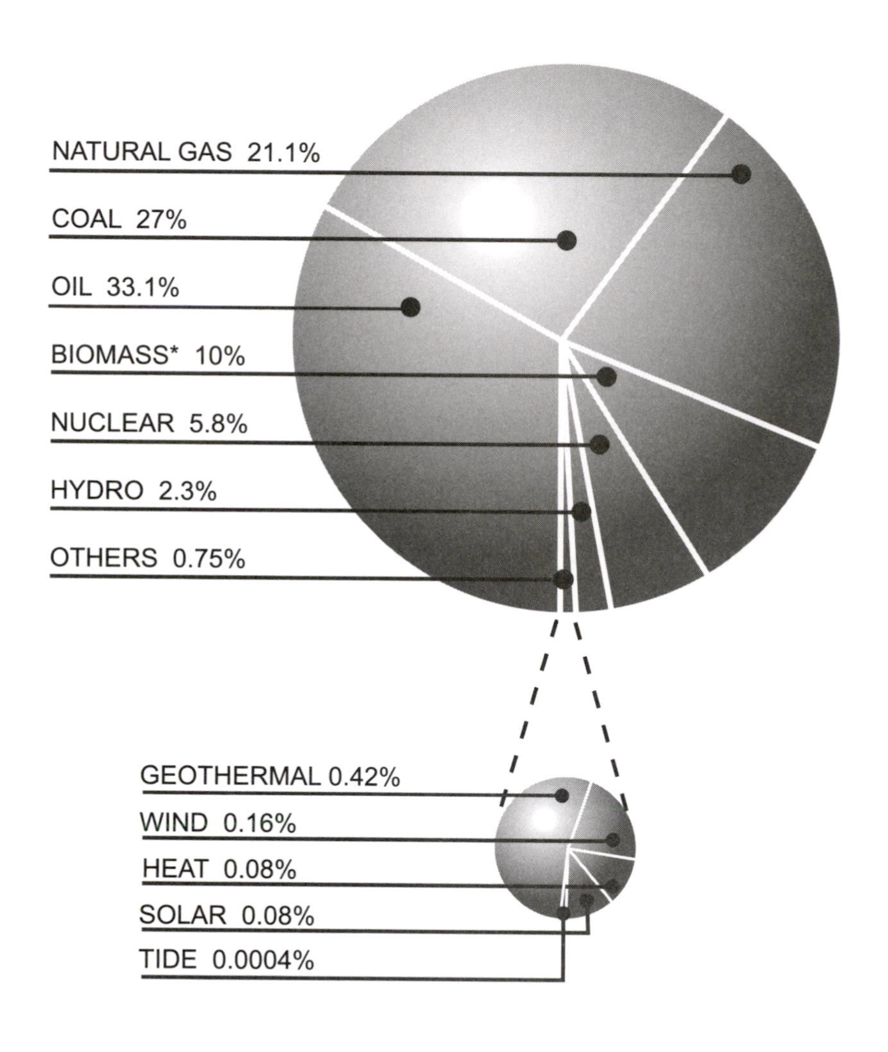

Figure 1 Global energy demand 2008 (primary sources)

indications that the human activity since the Industrial Revolution has hurt our planet, but this one is unequivocal.

Exactly how much suffering the Earth can endure is still very much in debate, but one thing is sure: a sick patient needs treatment. Even though we are uncertain about the development of the disease, we cannot wait for it to kill the patient.

Prudence urges us to look for a way out of this deep dependence on fossil fuels. The path is not easy. As shown clearly in Figure 1, no alternative source has managed to chip away at the primacy of fossil fuels. Only biomass has made any headway, but there is much misunderstanding about the impact of this category on the environment, the atmosphere, and the quality of human life.

The biomass shown in the figure is not the corn-based ethanol auto fuel that may come to mind. It consists mainly of wood, vegetable waste, dried manure, and other natural materials or garbage. Lacking more efficient alternatives, the poorest populations (mainly in Asia and Africa) continue to burn them massively. These leftovers from prehistoric times cannot provide a solution for our times.

Next on the list of contributors to the energy portfolio of humanity, we find nuclear power. This modern source of energy has been able to establish itself relatively quickly (in about fifty years), even though its initial costs are very high compared to fossil fuels. It is also a clean resource, at least in the sense that it does not produce harmful emissions, although its radioactive wastes require suitable handling. However, although nuclear power costs less today than fifty years ago, there are still doubts about its ability to provide a convincing response to our near-term energy problems. Over the last fifteen years, these doubts have caused the percentage contribution of nuclear power to the global energy requirement to decline.

The last resource to play a significant role in the energy balance of our planet is hydroelectric power, the flow of water moving turbines to generate electricity. As a source, it is clean, renewable, and under certain conditions very economical. There are still vast, exploitable, low-cost hydroelectric sources of energy throughout the world. However, in many cases, their apparent availability is only theoretical. In practice, environmental, geographic, and social reasons make only a small subset of them useable.

Two sources on which decision makers and the public have pinned much hope and attention recently are sun and wind. As shown in Figure 1, their current contribution is almost irrelevant.

Table 1. World production of electricity by source (2008)

Source	*Production (TWh)*	*Share (%)*
Coal	8,243	41.0
Natural gas	4,277	21.3
Hydropower	3,216	16.0
Nuclear	2,734	13.6
Oil	1,084	5.4
Biomass & Waste	259	1.3
Wind	216	1.1
Geothermal	62	0.3
Solar	9	0.0
Tidal & Wave	1	0.0
Other	4	0.0
Total	**20,105**	**100.0**

Only slightly more significant than sun and wind is the role of geothermal energy. This is a renewable source of energy fed by the flow of heat rising from the depths of the Earth in the form of steam or hot water. Finally, the role played by the tides is negligible.

If we look only at the generation of electricity (the sector of choice for renewable energy), the picture does not change much. The numbers in Table 1 are self-explanatory; decimals had to be added to the percentages so that the smaller contributors would show up at all.

A word of warning is in order on the subject of electricity from renewable sources. The media, politicians, and self-styled experts often provide misleading numbers for the actual potential of many renewable sources by confusing *installed electrical capacity* with actual *electricity generated*. There is a world of difference between these two. For example, suppose there are two power stations next to each other with the same installed capacity, say, 50 megawatts (MW). The first is powered by natural gas; the second by photovoltaic solar cells or wind turbines. Although they have the same installed capacity, the natural gas plant (or an oil-, coal-, or nuclear-powered one) will produce about three times more electricity than the sun or wind plant, because the latter will produce electricity only when it is sunny or windy.

In spite of the attention it receives, hydrogen is not a primary energy source but a secondary one. It is an *energy vector* (i.e., it provides energy obtained from another energy source). In the near future, the most economical way to obtain hydrogen will be to "extract" it from fossil fuels (in particular, natural gas), which today provide 95 percent of the hydrogen used industrially in the world.

The current contribution of all renewable resources (wind, solar, geothermal, and tidal) except waterpower to world energy consumption is less than 1 percent. This discouraging datum immediately raises the question: "Why?" The usual answer is that they still cost too much, but this is not their only problem. To get at a comprehensive answer, we should focus on two factors that are often ignored or dismissed in the public debate: *energy density* and *power density*. These factors are largely responsible for the energy trap that is affecting our quality of life.

Energy density measures the amount of energy contained in one unit of fuel (usually expressed in megajoules per kilogram). It is closely connected to the other factor, power density, which measures the rate of energy production per unit of land area being considered (expressed in watts per square meter). Let me try to simplify the explanation of these concepts.

At our present state of technical development, only nuclear power can surpass fossil fuels in energy density and power density. This means that they can provide enormous quantities of energy from small amounts of raw materials and a small area. On the contrary, the other sources require enormous quantities of primary energy and vast areas to obtain modest amounts of usable energy for our daily lives. Professor Richard Muller provided one of the best examples of the importance of energy density when he wrote, "for the same weight, gasoline delivers 720 times the energy of a bullet"![1]

The advantages of fossil fuels do not end here. They are available when we need them, they can be bought and stored for future use, and they can be transported from one country to the other. Moreover, they are flexible enough to provide several forms of secondary energy, for example, transportation fuel, home heating oil, or electricity. This is not to mention their importance in petrochemicals, for which they constitute the benchmark raw material. Petroleum possesses these features to an unequaled degree, for which reason it has become the "king" of energy. None of the alternative sources of energy can touch it.

The first problem of the renewable resources is that they produce limited amounts of energy, at least with our present technology. Even if we accept their higher cost for social or environmental reasons, the fundamental problem remains. They cannot respond to the huge energy needs of humanity over the next two or three decades.

Solar energy represents a peculiar case. Although its power density is much lower than that of fossil fuels, its theoretical potential is huge. This is because the sun offers us an enormous quantity of energy every day, free of charge. The problem is that our present technology can exploit only a minimal fraction of this energy, and it is too expensive.

It is easy to believe that powerful fossil fuel lobbies have conspired to block the development of alternative sources, particularly the renewable ones. There is possibly no conspiratorial image more popular than that of evil cabals of multinational oil companies deciding the fates of peoples, leaders, and governments and, above all, blocking any chance of a rival source deposing the king of energy. Yet, however powerful and seductive, this image is completely unfounded.

For example, consider that the principal producer and consumer of hydrogen is the petroleum industry itself, or that petroleum companies are among the largest investors in and producers of solar electrical energy. Solar energy benefited from significant space-related investments by NASA and other American agencies, but the first Earth-based applications of solar cells were made possible by petroleum companies.

The search for alternatives to fossil fuels began in the early twentieth century. Yet we have not met the two objectives of making renewable sources both economical and capable of producing great quantities of energy.

Those seeking to place blame for the retarded development of renewable sources could look at the amount of research funding that the public sector devotes to fossil fuel alternatives. Over the last forty years, more than 90 percent of these funds in all industrialized countries went to nuclear research, dwarfing all other options. However, it would not be fair to blame the public sector, and this would still not explain why renewable sources have never gotten off the ground.

In reality, what has repeatedly worked against alternatives to fossil fuels is that for most of their history, fossil fuels have been too cheap to leave room for any other source of energy. Faced with unsatisfactory results from research into alternative sources, continued investment in them was simply not reasonable economically.

After an oil crisis, it is hard to remember that during the twentieth century the oil market (and coal and gas as well) featured contained costs and a constant tendency to overproduce. Overproduction alone would tend to drive prices down. There were very long stretches of weak market conditions between brief but intense crises. The most protracted crisis period occurred during the 1970s, with two *oil shocks* (serious economic-political turmoil caused by a sudden difficulty obtaining supplies). Those two crises were relatively short in historical terms (about nine years), although it felt like forever to those who lived through them.

In some ways, fossil sources of energy have provided their own environmental solutions over the course of history. The age of coal ended the systematic destruction of woods and forests throughout Europe. By the sixteenth century, deforestation had already left Great Britain essentially without timber. The age of oil made it possible to clean up the smog from coal in industrial cities. The advent of natural gas and the first nuclear power plants made it possible to reduce emissions and pollutants from coal as well as oil.

Knowing what we know today, the cleanup side of fossil fuel history gives us little comfort about tomorrow. Indeed, the combined effect of these three sources is now spreading a dark shadow over our future.

Environmental awareness and concern about climate change are recent developments in human history.[2] For thousands of years, humans have not cared about the collateral effects of their activities. We have ignorantly polluted and disposed of waste wherever it was convenient, and we have unwittingly produced highly toxic substances.

Criminal fraud by persons who knowingly continue to pollute has come about only since the end of World War II, as we gradually learned the effects of certain manufacturing methods and certain products. I often find myself talking to scientists and researchers who worked in chemical laboratories and oil fields in the 1960s and '70s, exposing themselves to all sorts of risks without any protection. This is similar to what happened during the first nuclear experiments. They told me that there was simply no awareness of the dangers involved. On the other hand, they found it euphoric to be involved in a pioneering enterprise.

Environmentalism and general awareness of environmental risks began in California in the 1960s. Then in the 1990s, the alarm was raised over the danger of climate change caused by human activities. Industrialized countries tried to adopt more or less severe environmental legislation, but this ran up against the overpowering growth of mass consumption, and the laws were weakened for fear of restricting economic development.

The new century opened with widespread awareness that we needed to change our consumption patterns and make them sustainable for the planet and for future generations. For the first time in history, there is a consensus that broadly defined principles of environmental and climate protection should take precedence over purely financial considerations. It is like shopping for food and choosing an item that we recognize as good for us instead of a cheaper item of doubtful origin.

Unfortunately, things are not that simple in the field of energy. Within about twenty years (by around 2030), worldwide energy consumption could grow by almost 50 percent, according to forecasts and simulations by the International Energy Agency (IEA), the principal international organization for energy forecasts and statistics. Forecasting energy market trends is an impossible art, however, and many IEA forecasts have been wrong. Nevertheless, I cite the IEA forecast because it is based on conservative assumptions, including a credible average annual growth rate in energy consumption of 1.6 percent from 2004 to 2030. In spite of the prudent nature of this forecast, the additional amount of energy required in only twenty years is impressive.

In my view, there is no danger of exhausting available reserves of petroleum or other fossil fuels in this century, as I will explain in the following chapters. Therefore, faced with this growth in demand, the real problem is how to limit the use of fossil fuels and find something cleaner to replace them. It will be a real challenge to solve this problem by relying (as many hope) on renewables, and not only because of their cost and limited energy and power density.

A critical point is that most of the growth in energy consumption in the world will be concentrated in poor or developing countries. Without energy, there is no economic development and no freedom from poverty. This need for energy tends to put environmental and climate concerns on the back burner. For the vast majority of these countries, access to low-cost energy will remain a top priority. It is no coincidence that rapidly developing economies such as China and India continue to use coal. These countries are also burning wood and even dried manure.

Furthermore, our present state of technical knowledge will not support a revolution of the ability of alternative sources to provide large amounts of energy soon. It is true that some renewable sources have shown promising results, but under limited, special conditions that cannot be replicated. In spite of the enthusiasm of lay observers, journalists, and hopeful researchers, small-scale experiments often prove to be impractical on an industrial scale. This happens in every technical and industrial field, and the energy sector is no exception.

Finally, there is the risk that prices for fossil fuels may remain too cheap. On one hand, this would cause even stronger demand for these sources, while on the other hand, there would be a continued drop in investment in new forms of energy. This has happened often in the past, and it happened again after oil prices fell in 2008. As a case in point, the greatest wave of investment in renewable sources took place during the oil shocks of the 1970s. However, when prices dropped, spending on alternatives to oil, gas, and coal evaporated.

The problem of too-cheap oil lies behind the behavior of advanced countries, those which could set the pace in both adopting more responsible consumption patterns and finding energy sources to replace fossil fuels. It is difficult enough to change the habits of Western consumers when oil prices are high; it becomes impossible when they fall. In general, an oil price below $50–$70 per barrel discourages investment in renewable sources and drastically reduces popular interest in energy saving and efficiency.

The environmental conscience now shown by many governments could help sustain interest and investment in alternative energy when fossil fuels are cheap. The new policies on the issues of energy, environment, and climate coming from the administration of U.S. President Barack Obama represent a breakthrough for the whole world in terms of political commitment toward a different energy future. Yet the risk remains that private investment may not keep up and that governments may rethink their commitments, as happened in various European countries after the economic turmoil of 2008–2009.

A special danger is that investors or the government may waste money on the wrong choices. The public arena has no shortage of rhetoric and slogans when debating about energy. We need to learn a lesson from the history of technologies that crashed on takeoff. The more that a prospective technology is hyped as being the long-awaited breakthrough, the more quickly it will be abandoned (sometimes unfairly) when it does not produce the expected results in the overly optimistic time frames that were initially touted. In other words, the media success of a technology can be the cause of its rapid descent into oblivion.

In part, this is caused by unwise speculation. After investing millions of dollars for years on esoteric projects, discouraged financial investors cash out as soon as they can sell the early prototypes to someone else, before those prototypes have proven their commercial value. In the United States, 95 percent of start-ups founded to develop energy innovations fold before reaching the market. Experts refer to this as the "Death Valley of commercialization."

Then there are the "green swindlers." In my experience, the world of energy innovation is swarming with them. These financial operators take the ideas of honest researchers, package them up, and sell them to the first careless investor that they can dupe. This starts the chain of disappointed investors backing unmarketable prototypes.

Besides making careless investors lose money, this chain of disappointment and suspicion skews decision making about the real possibilities of one option versus another. Sometimes, this backlash has penalized technologies that deserved to be pursued regardless of the cost or market conditions of a given point in time.

Research funding should not depend on the capricious celebrity of opinion makers, on speculators, or on media slogans. Technological progress needs steady support, because usually it takes sustained effort to achieve results.

The final problem with the energy trap that imprisons us is the huge infrastructure needed to move beyond fossil fuels. Public opinion makers generally underestimate this issue. It will take new energy grids, pipelines, and other distribution systems to bring new sources of energy to the market. They will be expensive, and they will have to compete with infrastructure created to support the use of oil, coal, and natural gas.

Many people love to compare a future energy revolution to the digital revolution that has so dramatically changed our lives in such a short time. However, the digital revolution gave us something completely new. The personal computer and the cellular telephone had no competitors to push aside. Future energy sources will have to compete with fossil fuels and displace the huge infrastructure that has been created for them.

As we will see on our brief trip through the energy world, the challenges before us are huge and require a long-term commitment. We must neither yield to euphoria, delusion, or complacency nor be distracted by the prophets of catastrophe.

The forces that drive world energy demand upward—demographic growth and economic development—are unstoppable, and it would not be a good thing to try to stop them.

I belong to that school of thought for which a post-Malthusian view of the world is totally wrong. I am not a "cornucopian," as doomsayers have labeled those belonging to this school. We simply look at facts and numbers, as revealed by history and science, without falling prey to unsupportable catastrophic predictions. For example, I am convinced—as are many others—that the course of demography follows the evolution of wealth and culture: a richer and more educated population inevitably results in a progressive decline of its demographic growth, and then stabilizes at a very low birth rate. This is not a hypothesis but the evidence that has repeatedly come from history. History tells us that *wealth and education have always turned out to be the best contraceptives, while misery and ignorance have inflated world population.* This is why I cannot be pessimistic about our demographic prospects. The more economic growth advances in the poorer or emerging parts of the world, the more their population expansion will slow and then reverse, as has happened in the West.

Furthermore, I do not think that economic growth will dangerously affect the endowment of resources of our planet. Their real size is still unknown, but is much larger than most expect, notwithstanding the falsehoods spread by doomsayers. Human ingenuity has always succeeded in stretching those resources through inventions and technologies that require fewer resources to get better goods and higher levels of wealth.

As masterfully summed up by Paul Romer:

> Human history teaches us . . . that economic growth springs from better recipes, not just from more cooking. New recipes generally produce fewer unpleasant side effects and generate more economic value per unit of raw material. . . . Every

> generation has perceived the limits to growth that finite resources and undesirable side effects would pose if no recipes or ideas were discovered. And every generation has underestimated the potential for finding new recipes and ideas. We consistently fail to grasp how many ideas remain to be discovered. The difficulty is the same we have with compounding: possibilities do not merely add up; they multiply.[3]

For these two reasons, I cannot be a catastrophist. But there *is* something that makes me uneasy. It is something that Pulitzer Prize winner Thomas Friedman pointed out, declaring that the problem of our age is not demographic growth per se. In the words of Friedman, the problem is that

> we will go from a world population in which maybe one billion people were living an *American* lifestyle to a world in which two or three billion people are living an American lifestyle or aspiring to do so. . . . The metric to watch is not the total number of people on the planet—it's the number of *Americans* on the planet. This is the key number and it has been steadily rising.[4]

In other words, the problem is, as *Foreign Policy* editor Moses Naim put it, "Can the world afford a middle class?" That is, can the world afford the greatest affluence in history, of people consuming like Americans and wasting like Americans? In Naim's analysis:

> While the total population of the planet will increase by about 1 billion people in the next 12 years, the ranks of the middle class will swell by as many as 1.8 billion. Of these new members of the middle class, 600 million will be in China. . . . By 2025, China will have the world's largest middle class, while India's will be 10 times larger than it is today.[5]

Faced with this unprecedented explosion of people who will demand the same (or similar) standards of living that only Westerners enjoyed until recently, the world's energy system will be put under an unprecedented stress. Therefore, while I remain fundamentally optimistic, I cannot hide my uneasiness.

The breadth and novelty of this situation make it the greatest challenge of this generation and future ones. As a consequence, using Romer's metaphor, we need not only better recipes to use our resources but also to accelerate the invention of those recipes and make them commercially available for the world as a whole.

It will take time to cultivate promising prospects that are still far from offering us a solution. The road may be long, but it is well marked. In the end, only a constant, intelligent, and conscious collective effort will lead us to a new energy era.

This effort must hinge on research. I believe deeply that science can solve almost any challenge of nature. We can achieve energy sustainability. I also admit that my most carefully reasoned hopes focus on solar energy, where I see a viable option for our future.

We cannot impose deadlines on either science or technology. It is true that researchers have often stumbled upon discoveries that changed the rules of the game, but we cannot plan on those. Instead, we can only cultivate our research efforts day after day, waiting for them to yield their fruit.

Meanwhile, for ethical if not for rational reasons, we must understand how absurd our consumption patterns have been. We have become accustomed to readily available natural resources at ridiculously low prices, especially when compared with the prices of many other luxuries that we enjoy. I am referring not only to energy but also to water and other resources, which we are used to having almost for free.

The uncomfortable truth that we must all accept is that cheap energy is not good for the health of our planet, and it is not compatible with the fight against climate change. People's quest for cheap energy has made fossil fuels the over-dominant actors of the contemporary world, it has made energy efficiency a subject of minor relevance, and it has depressed investment in new technologies to develop affordable primary sources of energy other than fossil fuels.

A real energy revolution can only start with an end to our desire for cheap energy, at least for a short part of our history. If we accept this principle, we may hope to have cleaner, affordable energy in two decades.

Definition of Resources and Reserves

Quantifying the natural resources that our planet offers is a complicated subject. Here are some essential definitions that I will use to frame the discussion.

- *Resources:* the overall physical inventory of a given mineral, with no reference to its economic value or the feasibility of extracting it. In other words, there could be a great quantity of a given resource that cannot be technically recovered or is not convenient to extract. For example, there is gold dispersed throughout the oceans and there are minimal amounts of uranium in the subsoil almost everywhere.
- *Recoverable reserves:* that portion of the resources that could be exploited with existing technology and could be recovered economically under current price conditions.
- *Proven* (or *certain*) *reserves:* that part of the reserves that can be produced normally using the existing extraction and transportation infrastructure.

It gets more complicated. In any mining sector, it is difficult to establish what resources are trapped in the subsoil or in surface mines. Thus, it is difficult to know the total amount of any specific mineral resource. Even after we have estimated the recoverable reserves, we may not be able to recover them immediately. The general principles that distinguish resources, recoverable reserves, and proven reserves are valid for all mineral sectors, but the details of the definition differ noticeably from one category of reserves to the next.

For example, for the hydrocarbon sector (oil and gas), the Society of Petroleum Engineers and the World Petroleum Congress define *reserves* as the amount of oil and gas contained in known deposits that can be estimated with reasonable certainty and are commercially recoverable under the economic, technical, operational, and regulatory conditions in existence at the time of the calculation or in the near future. They further break down this definition into three categories established by the meaning of "reasonable certainty":

- *Proven reserves:* profitable recovery of at least 90 percent of the reserves being considered
- *Probable reserves:* profitable recovery of 50 percent
- *Possible reserves:* profitable recovery of not less than 10 percent

These few elements highlight some important concepts.

First, the *reserves* (regardless of the mineral) are usually only a small fraction of the *resources*.

Second, resources and reserves increase over time, not because our planet produces new minerals but simply because our knowledge of the subsoil increases, thanks mainly to new technologies to probe the Earth more effectively. Technical evolution also makes recoverable what was not recoverable before, which increases the reserves without the discovery of new deposits.

Third, we must never forget that the classification of proven reserves depends on the price of a resource at a given point in history. For example, until a few years ago, proven reserves of crude oil were estimated assuming a price of $18 per barrel, and proven reserves of uranium were estimated using less than $40 per kilogram. Today the market prices for both are much higher, but proven reserves have never been recalculated.

In summary, the concept of reserves and resources is a dynamic one. It changes over time as a function of consumption, our knowledge of the subsoil, the evolution of technology, and the prices of the various resources at given points in time.

PART ONE

The Fossil Sources

CHAPTER 1

"King" Oil

During the 1970s, crude oil supported 50 percent of the world's consumption of energy. Today, that share has dropped below 34 percent, chipped away by coal, natural gas, and nuclear power. During the twentieth century, oil was used for everything, from transportation to heating, to electricity generation, and to plastics and synthetics. At least in the industrialized countries, however, its use is now concentrated in the transportation sector, the only area where it is truly irreplaceable. Nevertheless, petroleum still holds first place among energy sources, and it will probably keep that position for a long time.

Even as we use more oil for our energy needs than anything else, many use "black gold" as a scapegoat for evils, wars, conspiracies, pollution, and global warming. It is so "politically incorrect" to deal with oil that energy courses in some universities ignore it. What a sorry fate for a resource that has had such an impact on our contemporary world. Oil has trivialized the meaning of geographic distance through mass and individual transportation. It has molded our lifestyles and given us many of the comforts and advantages that we take for granted. Often without our being aware of it, petroleum is behind much of our daily life, for good or for ill.

Why has black gold assumed such an important role that we cannot free ourselves from it? The answer lies in the laws of chemistry, physics, and economics. No other energy source is simultaneously so powerful, ductile, adaptable, transportable, storable, and usable. Its relative cheapness for most of its history has allowed it to spread throughout the world.

The golden age of oil has been over for some time, but our growing use of it raises different questions for different users. Some worry about whether there is enough oil under the surface of the Earth. For others, the most pressing questions are more ethical than economic. The public is now aware that the life cycle of oil includes extraction, production, combustion, and disposal of residue, many more phases than we used to consider. Even if there is enough of it, can we accept its environmental and climatic impact? There is no doubt that the way we have used petroleum so far will leave a heavy burden on future generations.

Still others ask what will happen to oil prices, which are subject to extreme and unpredictable swings. Anxious consumers are not the only ones asking this question. Those seeking funding for research and technical innovation, those developing renewable energy sources, and those trying to achieve real energy savings also worry about the price of oil. Time and again, cheap oil has stymied their efforts.

I will try to answer all these questions and clarify some myths and realities about this raw material. Let me begin with a warning. Many may hate petroleum and blame it alone for the worst things that happen in our world. Even if we consider petroleum "the enemy," we cannot ignore it. For many decades, petroleum will continue to be central to any energy scenario. Furthermore, its cycles will influence the fate of all other sources of energy. Therefore, we cannot ignore oil while thinking about a new energy paradigm.

What Lies Below?

How much petroleum lies beneath the surface of the Earth? We do not know.

We do know that history is full of estimates that would be hilarious if they had not been taken so seriously. For example, during the 1920s, the Anglo-Persian Oil Company (British Petroleum or BP today) refused to buy a share in the area that in 1932 would become a sovereign state called Saudi Arabia, because the company estimated that the area did not contain a single drop of petroleum.

In 1919, the U.S. Geological Survey (USGS) predicted that the United States would run out of petroleum in nine years.[1] As those nine years went by, the price of crude oil rose, but so did the number of new discoveries of ever-larger deposits. Then in 1930, the Black Giant was discovered in Texas (which inspired the movie *Giant* with James Dean). It was the largest deposit ever found, and it almost destroyed the petroleum industry of the day.

During the 1970s, baleful predictions of dwindling petroleum reserves stubbornly returned. According to almost all the experts, universities, research centers, and even principal petroleum businesses, production would peak about the middle of the 1980s and then begin an inexorable decline. A CIA report ratified the "rapid exhaustion" of accessible deposits, and President Jimmy Carter warned that oil wells were going to "dry up all over the world."[2]

About this time, the Club of Rome exposed the limits of modern development. While the group deserves credit for having raised a real problem, its predictions were tainted by an exaggerated sense of catastrophe from data and simulations that would later undermine its scientific credibility. The experts of the Club of Rome predicted that oil production would peak in the middle of the 1980s and then drop to almost one-half by 2000. This new wave of fears was dramatically disproven. In 1986, oil prices crashed under a wave of overproduction, generated by new discoveries and the collapse of demand. To make the point, by 2000, crude oil production was 25 percent *higher* than in the mid-1980s. Recently the prophets of doom have returned, bringing with them a popular theory: "Peak Oil." American geologist M. King Hubbert originally formulated this theory in 1956. In simple terms, Hubbert maintained that by knowing the geological structure of a country, the amount of oil available, and its exploitation trends, it should be possible to predict crude oil production for that particular country.

The geologist displayed this idea using a symmetrical, bell-shaped curve across time, with ascending production on one side and descending production on the other. In 1956, he predicted that oil production in the lower forty-eight U.S. states would peak in 1972 and then decline. This prediction proved to be accurate (the peak arrived between 1970 and 1971), which brought fame, credibility, and an army of followers to its author. Since then, Hubbertians have tried to apply this model to the rest of the world, refusing to accept its basic limitations: Hubbert's curve is only useful if one knows with certainty the exact amount of petroleum resources and the technological developments that will make it possible to exploit them.

Hubbert was lucky in 1956, because the United States was the most explored and drilled region in the world. This knowledge made an accurate description possible. Nevertheless, even he was wrong about the decline of American productivity, which was slowed greatly by the use of new technology. However, Hubbert's big mistake was trying to apply his theory to the entire world. He guessed that world production would peak in the mid-1990s. His forecast of collapsing production inspired the experts of the Club of Rome.[3] Fortunately for us today, they were wrong.

Unfortunately for Hubbert and everyone else, the subsoil of our planet remains an unknown universe. We can only hazard vague estimates of the total amount of oil it contains. If we do not know the total amount of oil, we cannot determine future production trends. Not only are global predictions based on Hubbertian models unreliable, but they tend to create a media circus that feeds on sensationalism, not science.

Why is it so difficult to know how much oil there is down there? Why is it wrong to think that we are reaching the end? To answer these questions, we need to take a short trip into the mysteries of black gold.

Petroleum is the result of the death and decomposition of living organisms, covered for millennia by layers of impermeable rock. Trapped in porous rocks called *reservoirs*, they were subject to very high temperature and pressure, which transformed them into the energy sources that we find so precious today. There are no large underground caves or immense lakes of petroleum, only vast strata of porous rocks, with the oil and natural gas imprisoned in tiny pores that are mostly invisible to the naked eye. In fact, if you could see an oil reservoir, you would notice only a rocky structure seeming to have no room for oil—the way that pumice seems to have no room to absorb water. This is why it is so difficult to understand what is there and to measure it scientifically.

Only a small slice of the mineral resources that we estimate exist underground can be classified as proven reserves (including petroleum). Currently, the most credible studies estimate total global proven reserves of petroleum to be about 1.2 trillion barrels,[4] enough to satisfy current consumption needs for thirty-nine years. Adding the 1.4 trillion barrels of recoverable reserves extends the limit to eighty-six years. The figures on recoverable reserves exclude crude oil that costs more than $18 per barrel to extract. This dramatically underestimates the real total.

Total worldwide petroleum resources are known as *original oil in place.* The USGS calculates that they amount to about 8–9 trillion barrels, of which less than 1 trillion has been consumed.[5] In addition to familiar petroleum, there are the so-called *unconventional oils,* such as ultraheavy oils, tar sands, and shale oils, which the USGS estimates to total 8 trillion barrels, of which only 1.3 trillion barrels are deemed to be recoverable. These numbers point to an important reality. Worldwide petroleum resources are enormous, and proven reserves are only a small fraction of the overall total. Furthermore, vast areas of the planet have yet to be explored. Today, only one-third of the *sedimentary basins* of the world (the basins where hydrocarbons may be found) have been properly explored. Exploratory activity is based on the number of new *wildcats* (exploratory wells) drilled during a given period.

A few numbers can shed some light on this little-understood phenomenon. Between 1980 and 2006, about 70 percent of the exploratory activity worldwide took place in the United States and Canada, mature areas that contain only about 3 percent of the proven world reserves of crude oil. On the other hand, the Middle East was involved in barely 1 percent of the oil exploration, although it holds more than 65 percent of the proven reserves.[6]

Oil exploration trends are even more surprising if we look back through history. For example, since the 1930s only three hundred exploratory wells have been drilled in Saudi Arabia, compared to several hundred thousand in the United States. There has been even less exploration in Iraq, Iran, and Kuwait. To complicate the picture, we exploit only a fraction of what is known to be there. On average, no more than 35 percent of the oil in the known deposits of the world is currently being recovered with existing, cost-effective technology.[7] This means that most of the crude oil already discovered remains trapped in its underground prison. But it will not stay there forever.

Global statistics on all mineral resources (including petroleum) show that none of them shrank during the twentieth century, in spite of massive exploitation. Indeed, reserves of petroleum and other raw materials have constantly grown. New technology and favorable pricing have allowed us to discover the unknown and to extract the unreachable. As we will see in chapter 3, this phenomenon has also recently involved the massive development of U.S. shale gas, which a few years ago no one even considered in the picture of potential new supplies.

Although our planet is not a cornucopia of infinite resources, no one knows just how finite the resources of Earth are. We only know that many more resources exist, and that technology can make them available.

The "Cornucopia Effect" of Technology

Over the decades, new exploration and extraction technologies have greatly increased both our ability to explore the subsoil and the amount of petroleum that we can collect from it. This has brought good fortune to countries that adopted these sophisticated technologies, and limited the potential of those that did not. In areas of the world where advanced technology is used largely by private companies, as in the United States and the North Sea, the recovery rate exceeds 50 percent. By contrast, in most Arab countries, the Russian Federation, and other major oil-producing countries, it is less than 25 percent.

The use of advanced technology has revitalized petroleum basins that were headed for exhaustion. One of the most impressive examples is the Kern River Field in California, discovered in 1899. Initially, it was thought that only 10 percent of its heavy, viscous crude oil could be recovered. In 1942, after a cumulative production of 254 million barrels of oil, it was estimated that the field still contained 54 million barrels of recoverable petroleum. As Morris Adelman pointed out in 1995, "In the next forty-four years, it produced not 54 million barrels but 736 million barrels and it had another 970 million barrels remaining in 1986."[8] Yet even data reported by Adelman were underestimated. In November 2007, Chevron announced that cumulative production had reached 2 billion barrels. The Kern River is still producing more than 80,000 barrels per day, and in 2009, the state of California estimated its remaining reserves to be about 627 million barrels.[9] The explanation of this apparent miracle is the injection of steam into the subsoil, a technology that Chevron began using in the early 1960s.

To maximize petroleum recovery, the petroleum sector has developed a spectrum of techniques gathered into two categories: *improved oil recovery* (IOR) and *enhanced oil recovery* (EOR). Although some confusion surrounds the use of these two terms, in a general sense IOR technologies tend to improve the ability to recover part of the *mobile oil* left within a reservoir during the primary and secondary recovery stages, mainly by re-injecting water and natural gas into an oilfield. EOR technologies loosen the forces that tightly bind the oil within a reservoir and keep it from moving, thus making the oil recoverable. This is done using gases, chemicals, heat, and other options.

The combined use of IOR and EOR technologies can push the recovery rate above 60 percent. However, EOR is more expensive, so it was economically impractical when oil prices were as low as they were for most of the twentieth century. When prices began to rise, producers shopped individual options, judging the trade-offs between exploring for new deposits and using IOR alone. The periods of rising prices have not been long enough to generate the maximum recovery rates that technology can give us.

I believe that the availability of oil will not be a problem if advanced technologies for exploration and production are applied on a vast scale to old and new areas of the planet. Critics may argue that while there may actually be plenty of oil left underground, the "easy" and cheap oil has gone forever. This view is partially true, but it is also true that today's difficult oil will become tomorrow's easy oil, thanks to the economies of applying currently expensive technologies on a large scale. In the 1970s, North Sea oil was considered among the most difficult and expensive oil on our planet. Only a decade from its initial production, its costs had been halved.

Today the world consumes about 30 billion barrels per year, with an expected annual growth of less than 2 percent (how much less will be covered later). This means that there is enough oil for most of this century. Let me hazard a gutsy prediction: by the year 2030, more than 50 percent of the petroleum we know about today will be recoverable. By then, the amount of oil that we know about will also have grown, and it will be economically feasible to produce a significant slice of the unconventional oils with low environmental impact. These trends could bring the total recoverable reserves of crude oil to between 4.5 trillion and 5 trillion barrels.

However, this brings up two basic questions.

1. Will the price of oil support an era of advanced technologies?
2. Even if there is plenty of oil, should we give up the search for alternatives and continue to exploit oil as we have so far?

These two questions are closely connected. The price of oil not only determines the way that it will be used, but also the future of all other energy sources.

Boom and Bust: The Violent Cycles of the Price of Oil

Until they lost their preeminence in the 1970s, the Seven Sisters, the great Western multinational companies that dominated the oil world for most of the last century, came closest to being able to control the price of oil.[10] Yet even their success was only partial. Beginning in the 1950s, they tried to use secret agreements to curb world oil production to prevent having prices depressed by the uncontrolled development of Middle Eastern oil. However, by the 1960s, not even the Seven Sisters could contain overproduction and falling prices, the result of continued discoveries by independent companies.[11]

Revolt in the Arab world and the widespread nationalization of oil during the 1970s put an end to the domination of the Seven Sisters and ushered in the Organization of Petroleum Exporting Countries (OPEC).[12] The golden age of OPEC was short lived, however. Political and economic conflict among its members prevented serious discipline with regard to choices and objectives. In the 1980s, OPEC proved unable to manage the delicate balance of prices, demand, and production. In 1986, crude oil prices collapsed under a tidal wave of overproduction.

The market took over as the place where prices would be set. However, the market proved no more reliable than the Seven Sisters or OPEC in stabilizing prices and providing clear signals to its operators. Indeed, price volatility since the late 1990s has been worse than ever. Prices in real terms touched their lowest levels in late 1998 and early 1999, dropping below $10 per barrel for a few days. Then the most spectacular rebound in history culminated on July 11, 2008, at $147.29 a barrel, the highest intra-day trading level ever reached by oil. However, what counts for the record books is the settlement price at the closing of a trading session on the New York Mercantile Exchange. That record was set on July 3, 2008, at $145.29 a barrel. This was followed by the most spectacular collapse in history, with the price hurtling to $32 per barrel by December 2008, recovering to an annual average of $60 per barrel in 2009.

What happened? The history of petroleum has always featured cycles of boom and bust, with long stretches of falling prices and sudden rebounds. In some ways, this has always been the curse of petroleum. At different times, it has inspired attempts to control fluctuations through oligopolies and monopolies. It is no coincidence that the first great antitrust case in history targeted John D. Rockefeller's Standard Oil and broke it into thirteen pieces in 1913. The founding father of the modern petroleum industry, Rockefeller was also the first tenacious believer in the need to control it monopolistically to keep unforeseen fluctuations in crude oil prices from destroying the industry.[13] The Seven Sisters and then OPEC tried to achieve the same goal. They all failed. As tough as conditions may seem in our new century, the rise and fall of oil prices is nothing new.

The question remains: Why is the price of oil subject to such extreme and unpredictable variations? Rather than bore you with a theoretical dissertation, I will try to frame an answer by using the last boom-and-bust cycle. In so doing, I will try to highlight the similarities with past crises; we will probably see them repeated. Let us begin with an apparent paradox that has characterized every oil crisis and caused the crisis at the turn of the twenty-first century: The high cost that the world paid for oil until 2008 was the consequence of low prices, which for almost twenty years had discouraged the exploration and development of new deposits in the richest areas of crude oil on the planet.

From the middle of the 1980s until the new century, crude oil prices oscillated between $18 and $20 per barrel, depressed by excess supply that twice caused prices to drop below $10 per barrel (in 1986 and 1998–99). This reinforced the conviction that oil had become just another commodity. For those who believe that oil is a commodity like any other, there are still inflexible laws of economics that apply. If the production of a commodity remains greater than its consumption for very long, the price of the commodity will stay low until production capacity is reduced. Even before the 1980s, major oil-producing countries had reduced investment in exploration to a minimum. Although most of these countries were (and still are) working old deposits using outdated technology, there was still too much petroleum available in 2000.

Thus, it should be no surprise that known oil deposits were not developed, refineries were closed, and hundreds of thousands were laid off. Excess production capacity was reabsorbed, perhaps too much. When the demand for many basic goods took off again, dragged along mainly by emerging countries (especially China and India), production capacity appeared too limited to meet it. Prices began to rise suddenly, sometimes in markets without a particularly strong demand. Petroleum is *not* a commodity like any other. In the petroleum sector, the severe thinning of production capacity compared to demand is accompanied by another crucial factor: the near-elimination of *spare capacity* (unused production capacity). This is where the problems begin.

Spare Capacity: The "Mother" of Every Oil Cycle

In my opinion, spare capacity is the single most important element for understanding the cycles of oil prices and, consequently, the alternating fortunes of the other energy sources. Spare capacity is the amount of available supply that can be activated if normal petroleum supplies are interrupted. Stated thus, it may appear only to be a safety cushion against the unexpected, but it is much more than this. As David Hobbs wrote:

> Because of the long lead times for bringing on new supply or investing in the potential for reduced demand, spare capacity can mean visibility of future supply to meet any predicted level of demand over several years allowing for a major disruption. It should be no surprise that market nerves become frayed when the available long term spare capacity is less than the production of any single large oil exporter that is viewed as being at risk of total shutdown.[14]

Throughout the history of oil, reducing spare capacity to a minimum has driven prices upward suddenly. On the other hand, the existence of ample spare capacity has maintained low prices even during periods of serious political tension and war.

To understand the importance of a safety valve like spare capacity, it helps to remember that producing petroleum is not like making shoes or T-shirts or even like growing wheat. Bringing a large deposit into production can take eight to twelve years. If petroleum becomes scarce and there is no spare capacity, the world must share that scarcity for a long time, and oil prices climb. This rise in prices fosters a new cycle of investment, from which new production will flow. It also triggers gains in energy efficiency, consumer frugality, and the rise of alternative energy sources. By the time the new production arrives at the market, petroleum demand may have dropped.

This vicious cycle has been a feature of all oil crises of the past. It begins with a drastic reduction in spare capacity and ends with a petroleum overproduction crisis.

By way of example, let me take us back to the 1970s.

Between 1950 and 1970, the demand for petroleum exploded throughout the world, but production grew even more. The price of oil dropped to its lowest levels during the 1960s, in spite of attempts by the Seven Sisters to contain global overproduction. Meanwhile, bargain oil prices pushed consumption sky high and discouraged exploration and development.

In the early 1970s, demand continued to grow, but supply began to grow more slowly. In 1971, the United States, long the guarantor of spare capacity for the Western world, reached maximum available production. The market could sense that the era of petroleum security was over.[15] The sudden petroleum weakness of the United States allowed Arab countries to unsheathe the oil weapon against the West.

The West's worst nightmare materialized when the Organization of Arab Petroleum Exporting Countries (OAPEC, not to be confused with OPEC) placed an embargo on petroleum exports after the Yom Kippur War in October 1973. The embargo was limited and relatively little petroleum was taken off the market,[16] but the decision caused oil prices to explode. Between March and December of 1973, the reference price for OPEC crude quadrupled, reaching $11.65 a barrel.

Later studies confirmed that the oil shock of 1973 was a true panic,[17] caused by the absence of reliable data on the availability of petroleum and generalized fear that world reserves were drying up. In reality, there was enough oil, but the perception of insufficient spare capacity pushed the world market into shock.

Now let us return to our times.

Historical evidence indicates that spare capacity of less than 4 percent of world consumption can trigger a dramatic increase in prices. Spare capacity between 5 percent and 6 percent of world demand provides a cushion capable of keeping prices under control, while prolonged spare capacity of more than 6 percent can start eroding prices, especially when demand is stagnant.

By 2003, spare capacity had dropped to a very thin 2 percent of world consumption. Thus, minimal spare capacity made the price of oil dangerously vulnerable to almost any event: regional conflicts, hurricanes, pseudoscientific theories about the end of oil, market rumors, or financial speculation.

Spare capacity puts its distinctive mark on market expectations, a negative sign when spare capacity is too low and a positive sign when it is too high. If the market is affected by other factors, the consequences can be devastating. What are these factors?

Irrational Expectations, Poor Data, and Myopia about Oil Demand

In every major oil price cycle, at least four factors have worked with spare capacity to turn a tense situation into a full crisis:

- The perception of an imminent decline in world reserves, with a consequent drop in production
- Unreliability of data about actual supply, demand, and inventories, leading to a misrepresentation of the actual trends
- Overestimation of future oil demand growth
- World geopolitical relationships, especially concerning the major oil-producing countries

To these we should add a specific technical factor: petroleum refining, which I will deal with separately.

I have already touched on the first element, recalling how the idea of the end of oil has accompanied the modern petroleum industry throughout its history. This fear has reemerged with each oil crisis.

In the 1920s, the expectation that American crude oil was about to run out made prices almost triple in a few years. It also caused a surge of petroleum imperialism by the Great Powers, each wanting to get its hands on petroleum in other parts of the world.

During the 1970s and in the last ten years, various versions of the Peak Oil theory made their impression on the collective psyche. Each crisis brought a new wave of petroleum doomsaying. If the market believes that the global petroleum production is about to drop, it will conclude that spare capacity cannot be replenished. If oil demand keeps growing, crude oil prices begin to rise without stopping.

Unfortunately, it is too easy to forget the past mistaken predictions of the prophets of Peak Oil. For example, their principal exponent in our days, Colin Campbell, has revised his estimates of ultimate recoverable resources several times (in 1989, 1990, 1995, 1996, and 2002), always upward. When his predictions proved wrong, Campbell simply moved the peak point farther out in time.[18]

The second factor, the poor quality of petroleum data, has tormented the petroleum market since its beginnings, playing a fundamental role in the panics of 1973 and 1979. Unfortunately, we still do not have reliable data. For example, China and many developing countries have never adopted a complete system for collecting petroleum statistics.

Many advanced countries provide consumption and inventory data slowly, so OPEC must use secondary sources to assess the actual production of its own members. Accurate data on consumption, production, and inventories emerge with a delay of one or even two years. This means that the figures from last year or last month are only estimates, likely to be proven wrong next year.[19] Having to rely on unsatisfactory data makes current forecasts about demand, consumption, inventory, and prices grossly inaccurate. In 2001, the International Energy Agency (IEA), OPEC, and several other organizations launched the Joint Oil Data Initiative (JODI), but it has not produced significant results to date.

Forecasting and estimating errors have plagued even the most important source in the world for petroleum data, the IEA. The agency is the principal source of data and analyses supporting governments, financial institutions, oil companies, and even oil-producing countries. Moreover, its forecasts influence the expectations of analysts and market operators. It played a questionable role in the first oil crisis of this century. Beginning in 2005, the IEA routinely overestimated world demand for petroleum, creating a perception that oil production could not keep pace with consumption.[20]

The IEA was not the only one making bad predictions. During the same period, almost everybody was wrong, and all in the same direction. This exaggerated demand forecast created a market perception of demand shock. In 2004, the average price of crude oil hovered around $38 per barrel. It climbed to an average of $54 in

2005, $65 in 2006, $72.50 in 2007, and $93 in 2008 (after touching points above $140 that year).

In reality, between 2001 and 2007, world demand for petroleum grew by a modest average annual rate of 1.7 percent, barely higher than it did between 1991 and 2000 (1.4 percent), which was generally considered a decade of sluggish demand. The slightly higher demand, especially after 2003, was in part a reflection of strong global economic growth after many years of relative stagnation, but it was also influenced by two factors.

The first was easy credit in many advanced economies. This allowed families and businesses to maintain exceptionally high levels of consumption, which would have been unsustainable otherwise. The second was the manipulation of energy prices by governments in many developing countries through subsidies, beginning with China. Artificially low retail prices for gasoline and other petroleum products encouraged consumption.[21]

Furthermore, the average rate of the increase in demand during this period was distorted by two specific years, 2003 and 2004. In 2003, the growth rate was 2.1 percent (equal to 1.6 million barrels per day [mbd]). In 2004, demand reached a peak of 3.7 percent (about 3 mbd), the highest annual increase since the 1970s. This was the closest thing to a "demand shock" recorded between 2001 and 2007.

The next three years, petroleum demand grew even more slowly than in the 1990s (1.7 percent in 2005, 1.3 percent in 2006, and 1.1 percent in 2007). At the same time, except for 2007, oil production increased in absolute terms faster than demand. Therefore, it is inappropriate to describe the crisis of the early twenty-first century as demand driven. Rather, it was driven by mistaken expectations about demand trends.

The same assessment error was committed during the 1970s. Because crude oil consumption grew at about 7 percent during the period 1950–1970, growth in demand was overestimated by extrapolating future trends based on past statistics. During that period, even the most conservative predictions assumed growth rates between 4 percent and 5 percent for the 1980s and '90s.

They were all wrong. From 1980 until 2000, the demand for oil grew at less than 2 percent per year, while the growth in production was much higher.

Why this sequence of errors on predictions of demand? Probably because most analysts were (and remain) trapped by three assumptions:

1. Future economic growth will translate into growing demand for petroleum.
2. World demographic growth will push up oil (and energy) consumption.
3. Present and past statistics can be used to extrapolate future growth rates in demand on a straight line.

The traps in these assumptions can be summarized thus:

- Technology and energy efficiency have consistently reduced the oil input for each unit of GDP. In other words, it takes less oil today to create one additional U.S. dollar of new wealth than in the past. In particular, each oil crisis has led to a jump in efficiency that has reduced the specific consumption of petroleum for each dollar of wealth created.
- Like the demand for petroleum, the statistics of future demographic trends always seem to feature unstoppable growth. In reality, developed countries have fewer children and a lower specific consumption of energy for each unit of wealth created, because they can take advantage of new technology and more efficient energy systems.
- The statistics of the past tell us nothing about future technology or efficiency improvements. Therefore, it is deceiving to use past data to predict the future.

Finally, no one knows the limits of petroleum reserves, but we do know that what is available is more than abundant.

In other words, the future cannot be projected based on linear models of the past, in which only demand rises while supply falls. This is the most common mistake that these forecasters have always made. They will probably continue to make it.

Unfortunately, an irrational misperception of the future may be a stronger force than spare capacity in determining the course of oil prices. In the early months of 2009, for instance, the oil prices rebounded from their lows of December 2008 and exceeded $70 per barrel, just when spare capacity was rapidly increasing and oil demand was moving downward because of the economic crisis. There was only one explanation for that situation: the market remained convinced that once the economic crisis passed oil consumption would jump, again putting world crude production capacity under stress.

Oil Prices and Geopolitics

Almost 65 percent of conventional proven reserves of crude oil are held by five countries around the Persian Gulf: Saudi Arabia, Iran, Iraq, Kuwait, and the United Arab Emirates. Outside the Gulf region, only two countries currently have large proven reserves: Venezuela and the Russian Federation.

In most countries, petroleum reserves are controlled directly by the state, usually through state-owned companies. Conversely, the percentage of proven reserves held by private companies is only about 8 percent of the world total. The Seven Sisters no longer exist. The top seven among their shrunken heirs, called today "Super-Majors" or "International Integrated Majors," control only 4 percent. In order of size, they are: ExxonMobil, BP, Shell, ChevronTexaco, Totalfina, Conoco-Phillips, and Eni.

It is easy to see why the interaction of the geopolitics of petroleum with spare capacity is so important. Actual production of crude oil is more widespread than the location of proven reserves, but at least 50 percent of crude oil production comes from countries considered critical or unstable. If spare capacity drops while the winds of war blow across major oil-producing countries, the market will fear that the war will interrupt production and prices will climb. On the other hand, if war breaks out when there is elevated spare capacity, operators and analysts will remain relatively indifferent to geopolitical problems, because they know that there is a production reserve ready to go to market if needed.

For example, during the 1980s eight years of war between two of the world's largest oil exporters, Iran and Iraq, did not stop the fall of crude oil prices. In January 1991, on the day that United Nations forces attacked Iraq (following the invasion of Kuwait), the price of oil dropped from $30 to $20 per barrel. In both cases, there was ample spare capacity.

Things went the other way in this century. Fear of an American attack on Iran, the recurring threat of a Turkish invasion of northern Iraq, the unending instability in Iraq, and near civil war in the Niger Delta pounded a market in which spare capacity was too low. This aggravated the expectation of price increases for petroleum.

Fear of a politically motivated curb on oil supplies has always been part of the energy debate in the West, apparently legitimating calls for the West to secure its own oil independence. This fear has a long history. Referring to British behavior during the 1956 Suez Crisis, historian Hugh Thomas observed, "Ever since Churchill converted the Navy to the use of oil in 1911, British politicians have seemed indeed to have had a phobia about oil supplies being cut-off, comparable to the fear of castration."[22]

This castration syndrome in consuming countries (both advanced and rising economies) has often had a mirror image: resource nationalism among oil producers.

However, a shortage of crude cannot last forever. The extensive segmentation and flexibility of the market and competition between producers keep that from happening. The impossibility of specific forces completely controlling oil supply and prices was proven when the world oil market featured powerful oligopolies.

The Seven Sisters were only partially successful in restricting production to sustain prices from 1950 to 1970, because they could not escape the competition from independent companies or the reactions of producing countries. OPEC did not plan the 1973 shock and could not manage its apparent windfall. The cartel merely exploited consumer anxiety, boosting oil prices higher while supply was widespread and growing. The result of that mismanagement was the countershock of 1986.

OPEC today controls little more than 30 million barrels per day of the roughly 84 million barrels that the world consumes daily. The organization is hardly monolithic. Its members have different ideologies, policies, and economic targets, governed more by self-interest than by a sense of common purpose. Meanwhile, international oil companies need to replace reserves every year to sustain future production, so their main interest is to open new frontiers and experiment with new technologies in their quest for survival. The more that OPEC limits production to raise oil prices, the more incentive it gives international oil companies to develop new resources outside the cartel.

As we have seen from another point of view, extended shortages breed high prices, which in turn spur investment in higher-cost production areas. Oil can be a weapon, but only for brief periods. As a weapon, it is prone to backfiring, as the Arab oil producers learned the hard way in the 1980s.

Historically, consumer and producer countries have sometimes come close to clashing, with consumers struggling to retain access to a key resource and producers trying to use it for leverage. In general, the laws of the marketplace, though imperfect, have usually prevailed, in spite of the emotional images in the collective imagination.

However, explanations rooted in economics or historical evidence have never displaced the myth of oil insecurity. Public perception of oil issues today is still colored by the oil shocks of the 1970s and the selective Arab embargo in 1973. These incidents skew any discussion of oil, despite the fact that they were largely the product of a distorted collective psychology. At that time, catastrophic predictions spread by the media and think tanks and misguided interventionist policies by the United States only confused the issue further. It is worth recalling that the real oil shortage was small and manageable. In any case, those particular shocks were exceptional events in the history of oil.

In fact, the major oil producers learned a powerful lesson from the 1970s. An oil shock can be a terrible experience for the industrial countries, but it is not a fatal blow. As soon as the consumer countries perceive that the shock will endure, they react, and their reaction can become a producer's nightmare. Structural reactions lead consumer countries not only to slash the rate of growth of demand but also to increase investment in research, the development of alternative sources of energy, and the exploration of new oil-producing areas.

Whatever their problems and in spite of uninformed perceptions in the West, Iran and its Arab neighbors have always made every possible effort to ensure the stability of oil prices and supplies, and they are still trying to pursue this objective. This does not mean that one of them will not be tempted to use oil as a weapon, especially when spare capacity is low and both prices and tensions are high. However, this scenario would be an exception, not the rule; most importantly, it could not last for long.

In sum, oil is both a blessing and a curse for those countries that depend heavily on its revenues. They are more like Dionysius than Damocles, because they cannot simply step out from under the sword and leave. A selected embargo against "enemy countries" would hardly affect the global oil market; countries not on the blacklist could always resell oil to the embargoed countries. As Adelman put it, "Whether a supplier loves or hates a customer (or vice versa) does not matter because, in the world oil market, a seller cannot isolate any customer and a buyer cannot isolate any supplier."[23]

The Problem of Transforming Petroleum

The organization of the refining sector is a special problem that has played a more important role in the crisis of this century than in earlier ones. It probably will resurface sooner or later. Because of its complexity, we need to make a small digression.

Petroleum is not a homogenous commodity. It comes in many types and qualities. Its two most important characteristics are density and sulfur content. Higher-quality petroleum features low density (light crudes) and low sulfur content and commands a higher price. Examples include the reference crude oils of the international market: West Texas Intermediate (WTI) and the British Brent.

Each type of petroleum provides different amounts of final products such as gasoline, diesel, jet fuel, and heating oil, collectively known as its *yield.* The higher the quality of the crude oil, the more high-value product (e.g., gasoline and diesel) can be obtained by a *simple refinery* (equipped only with a primary distillation unit).

A simple refinery cannot significantly change the standard yield of a given quality of petroleum, and it is always left with a residue containing a high percentage of heavy fuel oil and sulfur. The percentage increases or decreases depending on the type of crude oil processed by the refinery. Light crudes leave less residue; heavier crudes leave more.

To handle a wider spectrum of crude oils and obtain a better combination of high-value products, a refinery needs a *deep conversion* installation, which drastically reduces the amount of residue. It also needs a *desulfurization unit* to remove the sulfur. Refineries with these installations are more sophisticated, complex, and expensive.

This brings up a point that is not well understood by the public: Refiners do not simply need "petroleum," but particular *types* of petroleum suited to their installations, with which they can satisfy the specific demand for petroleum products in their markets. This can create an apparent market paradox. The world can be swimming in oil, but the refineries may not be able to find the type of crude oil that they need.

Three factors converged during the first decade of this century. First, the worldwide refining system had been impoverished by two decades of plant closures and inadequate investment; deep conversion units were in especially short supply. Second, the world wanted lighter products with better environmental qualities, while the demand for heavy fuel oil was falling. Third, after 2004, much of the growth in the petroleum supply consisted of medium-heavy crudes with high sulfur content. These came to represent about 80 percent of the worldwide petroleum supply.

This combination took shape in 2000 and 2001, but became critical after 2004. Refineries entered a phase of stiff competition for the specific grades of crude oil needed to feed their systems. Paradoxically, some crude oils remained without a market unless steeply discounted. Others skyrocketed because of high demand. All this increased the stress on a market that was already in crisis.

Why Do Oil Bubbles Burst?

We have seen why the price of oil can begin to rise or fall, but we still do not know how far it will go in either direction. The lack of methodologies to explain the magnitude of rises and falls in the price of oil obscures the similarity between the two. It reminds me of a comic strip in which two characters come down a steep hill, look back, and one says to the other, "Funny how downhill looks like uphill when looking up from down."

The similarity may indicate that the price will go up as steeply and as irrationally as it will later fall. However, there is nothing scientific about this observation. It is based only on empirical observations from which emerge some general indications. I will try to summarize them.

Increases in investment do not instantly follow oil price rises. Usually, there is a time lag before investors become convinced that the increase is sustainable for a long time. But as soon as they do, investments take off and new production capacity is developed wherever it is profitable. As we have seen, it takes time for the oil to arrive to market, so prices continue to rise.

However, as prices rise, they slow the growth of the economy and the consumption of petroleum. Thus, rising oil prices create the conditions both to increase production and to reduce demand. When these two converge, falling prices are around the corner. The warning sign for change is spare capacity. A solid expansion in spare capacity signals that world production capacity is overtaking the growth in demand. Sooner or later, we can expect a wave of petroleum overproduction.

Every petroleum price cycle in history has featured these ascending and descending curves. For example, between the 1970s and the 1980s, astronomical prices for crude oil led to the development of the North Sea and Alaska. They also gave a noticeable push to investment in traditional areas, especially Mexico and the Soviet Union.

The effects of petroleum investment after 1973 were felt in the early 1980s, when almost 6 mbd of new oil hit the market, as worldwide consumption was falling (from a demand peak of 64 mbd in 1979 to about 58 mbd in 1983). Demand was contracting partly because of high prices and partly because of laws passed by many governments to impose greater savings and energy efficiency. In 1981, the price of oil began to fall. From a historic high of $42 per barrel in 1980 (and an average of $36 per barrel for that year), prices fell to below $10 a barrel in early 1986.

In 2008, crude oil prices collapsed because the world economic crisis drastically cut the demand for petroleum, but demand had already been slowing in many parts of the world. It had been contracting in the United States since 2007, and in Europe and Japan since 2006. By 2008, demand was growing only in developing countries, mainly because of government subsidies that kept the retail price for oil products below the international market price.

Even without the financial-economic crisis in September 2008, sooner or later prices would have fallen. They might have fallen less drastically than they did, perhaps over a longer period (as in the 1980s), but they would have fallen.

Meanwhile, supply continued to increase. In November 2008, oil production set a historic record, almost 88 mbd, with 3 mbd in spare capacity.

Unconsumed product filled inventories, often becoming invisible, because lagging international data collection could only report a statistical discrepancy between supply and demand (a phenomenon known as *missing barrels*). Even in March 2009, spare capacity was estimated to be close to 7 mbd, about 8 percent of world petroleum demand. As in the past, it seemed that no one noticed that the price of oil had stopped rising, except for the occasional prophet crying in the wilderness.

It is typical of each phase of climbing oil prices that people get used to thinking that something has changed forever and that expensive oil has become an inescapable fact of life. The same thing happens when prices fall; the public becomes convinced that the price will keep going down. Let me share two anecdotes on this subject.

In March 1999, oil prices were slowly recovering from the collapse that had pushed them below $10 a barrel at times. That same month, *The Economist* dedicated a lengthy analysis to the future of petroleum, predicting that a long-term price of $10 a barrel would prove to be optimistic and that black gold could even drop to $5 per barrel.[24] At the time, the prestigious British magazine was almost unique in observing that such a low price would have negative consequences. For the most part, Western observers seemed very happy with the prospect of cheap oil. Yet that very month, petroleum was silently starting its upward path, and it would not stop until it reached $147 per barrel in July 2008.

Almost ten years later, *Fortune* published an article entitled "Here Comes $500 Oil," celebrating the predictions of one of the best-known doomsayers of Peak Oil, Matthew Simmons.[25] His forecast was only the culmination of what most analysts had been predicting: that the price would keep increasing. Yet just one week before the *Fortune* article, the Lehman Brothers bankruptcy was announced, and the price of oil plunged into the sharpest collapse in its history.

The Economist in 1990 and *Fortune* in 2008 reflected views largely shared by policy makers and those who shape public opinion. A list of them would be too long and certainly somewhat embarrassing.

What Awaits Us down the Road?

Today, conventional wisdom holds that a fall of prices should be short-lived and that its negative effects will be delayed. It will cause a reduction in investment needed to create new production capacity. When world consumption takes off, petroleum supplies will lag, and the price of oil will soar again.

I was among the first to express this opinion. In 2006, writing for *Foreign Affairs,* I warned that rising prices for crude oil would sooner or later cause a new fall in demand and thus in prices. This was a contrarian position when almost everyone believed that prices could only increase, indefinitely. I maintained that a sudden fall in prices would stop the boom in investments then under way and lead to another crisis tied to supply deficiencies.[26]

I still believe that this position is correct and that the future holds more boom-and-bust cycles for us. Before hazarding bold predictions about price, we must carefully consider three crucial factors.

First, we must determine the level at which prices will foster the investment needed to create adequate production capacity.

Most experts believe that there is a natural floor for the price of oil, known as the *marginal barrel*. This is the most expensive barrel to produce that will satisfy the last barrel of global demand. Even in early 2009, many experts maintained that the natural long-term price would hover around $75. However, at current and foreseeable demand, the most expensive barrel needed by the market costs about $50, and this price could drop gradually as the inflation of the first decade of this century subsides (see sidebar).

No doubt, easy and cheap oil seems bound to disappear. Between 2003 and 2008, exploration and production costs more than doubled, driven by increases in the cost of steel, engineering and construction services, and skilled labor.[27] The *government take* (taxes and royalties) imposed by producing countries escalated over the same period, increasing the overall cost of delivering a barrel of oil to the market. Finally, operating on new petroleum frontiers like the Arctic Ocean or the ultradeep offshore rigs off Brazil or in the Gulf of Mexico will initially be more expensive than $75 per barrel. Revitalizing production in the older and bigger petroleum basins of the world will cost more, too.

As I explained before, today's cheap oil was not that cheap and easy when it was first discovered and developed. Over the long term, the learning curve associated with developing new, difficult areas and using advanced technologies will move downward, as will the cost. The government take of the producers will depend not only on the investments they will be able to attract over the coming years, but also on the competitiveness of their oil on the final market. Furthermore, as Ed Morse has stressed, since 2008 "the costs of production have fallen faster than capital expenditures," while many rigs and vessels for oil activities were under construction—auguring a situation of market oversupply in the next few years.[28]

As a consequence, it should be advisable not to bet on high prices for oil that were simply unbelievable at the end of the last century, when most oil companies, producing countries, and investors considered anything over $25 per barrel to be high.

How Much Does It Cost to Produce One Barrel of Petroleum?

The unit cost of production is known in the petroleum trade as the *technical cost*. It is simply the total cost of production divided by the producible reserves (the total amount that can be extracted from the subsoil), but it is affected by many variables. These geological, technical, and economic variables span the entire life cycle of the deposit, including discovery, development, extraction, maintenance, and reclamation of the abandoned site.

Each deposit has unique characteristics. The technical cost varies not only between different countries but also within the same country and within the same deposit. For example, the largest deposit in the world, Ghawar in Saudi Arabia, extends over an area as big as the state of Delaware and contains six distinct mining areas.

This diversity requires different technical solutions to recover the maximum possible amount of petroleum from the subsoil. In many Middle Eastern deposits, the crude oil rises to the surface essentially by itself, so simple technologies are sufficient. In other cases, complex processes and installations are needed to extract, treat, and transport the petroleum and everything that is necessary to produce it.

In general, the ranges of technical cost of petroleum are:

- $2–$5 per barrel in the deserts of the Middle East and North Africa
- $5–$7 per barrel in the Venezuelan forest
- $7–$12 per barrel in the tundra of Siberia
- $25–$30 per barrel in the frozen Arctic of Alaska
- $20–$25 per barrel in the shallow waters of the Caspian Sea
- $25–$30 per barrel in the ocean depths off Angola
- $35 per barrel in the Barents Sea and the bituminous sands of Canada

Naturally, even in areas like the Middle East, new projects could be more expensive, because the easy oil has already been discovered and put into production.

The technical cost serves as a minimum threshold for the price of petroleum. In addition, other elements must be considered, which significantly affect the final cost of a barrel on the market: taxes and royalties paid to the producing country, logistical costs, and a return of at least 10–12 percent for the operators.

Finally, the costs vary not only across space but also across time, depending on market conditions; the learning curve in developing a deposit; and the evolution of technology.

Second, we must consider the future demand for petroleum.

This will depend on the level of global economic growth, particularly in emerging countries. However, we must analyze carefully the relationship between economic growth and the demand for petroleum, to avoid making the mistake of thinking that after each crisis everything will go back to the way it was.

History has shown that industrial societies and emerging countries both tend to introduce structural changes following an energy crisis, modifying their own consumption patterns and improving energy efficiency. Per-capita consumption of petroleum and its impact on wealth formation in the industrialized countries is much lower today than it was in the 1970s because of laws passed to encourage savings and energy efficiency. Europe, Japan, and Australia recorded a peak in petroleum consumption during the 1990s, but since then they have substantially reduced their consumption.

The United States is the only advanced country in which the demand for petroleum has increased constantly since 1984, and it remains the most inefficient energy-consuming country in the developed world. Before the current economic crisis, the country's yearly per-capita consumption of oil was twenty-six barrels, compared to twelve barrels in Western Europe.

Nevertheless, the era of unrestrained consumption may be coming to an end in the United States as well. A new attitude toward energy, climate change, and environmental pollution seems to be moving the course of American politics. Much will depend on the severity of the policies undertaken by the Obama administration.

If the United States can commit to a serious program for energy efficiency (see chapter 12), one could expect that within ten years the country could reduce its own consumption by 20 percent (all the other variables remaining the same), that is, by 4 mbd. It should not escape notice that this would be equal to the total current production of Iran.

Meanwhile, the European Policy on Energy (EPE) was established in December 2008 with the evocative slogan "20-20-20 by 2020," indicating its three objectives: 20 percent reduction of energy intensity; 20 percent reduction in carbon dioxide emissions, and a 20 percent increase in the use of renewable energy. Even if 20-20-20 seems a little unrealistic, the EPE could further cut European consumption, which was already dropping.

The demand for oil will continue to grow only in developing countries, with China and India leading all others. We need to understand how these countries and others like them can improve their energy efficiency, "generating" more energy by consuming it more effectively.

Currently, China uses energy inefficiently and has become the biggest emitter of CO_2 in the world. A large part of the infrastructure in China dates back to an era when no one paid attention to energy efficiency. It was only around 2000 that this became a pressing problem for planning new electrical power stations, transportation systems, and even urban planning. Considering all the elements that I have tried to summarize, I would not be surprised if the increase in global demand proves to be more contained than expected, even less than 1 percent per year on average over the next twenty years. The key to the evolution of future demand will be the balance between declining demand in industrialized countries and rising demand in emerging economies.

Third, we cannot exclude the oil-producing countries and their legitimate concerns from our calculations.

While the industrialized world is concerned with the safety of petroleum supplies and obtaining crude oil at a bargain, the producers are concerned with the safety of demand at a price that will compensate their investments. If producers come to expect that the thirst for oil will diminish, why should they bother investing to increase production? If OPEC countries were to curtail their investment programs, the crude oil supply would contract, leading to several years of price increases. In other words, it would be a serious error to ignore the legitimate concerns of oil-producing countries, because their reaction can aggravate the volatility of the petroleum market.

The Role of Speculation

Adding the financial dimension complicates the petroleum picture even more. The petroleum financial market uses *paper barrels,* which are financial securities (futures) tied to petroleum. Contracts exchanged on these markets are for lots of 1,000 barrels per contract. They do not involve the physical delivery of any crude oil; they are simply financial transactions.

Since the second half of the 1990s, the exchange of paper barrels has increased exponentially, and they now outnumber the exchanges of physical barrels of petroleum. The presence of purely *financial operators* (who are only speculating on the rise and fall of the price) in these exchanges has led many observers to conclude that speculation now plays a fundamental role in determining crude oil prices.

Let us look at how this market works. As with other raw materials, the principal raison d'être of the petroleum financial market is to offer *commercial operators* (oil companies, airlines, refineries, etc.) a vehicle for managing risk. The most significant risk is price volatility.

Although commercial operators still represent the largest segment of the market, the financial operators cannot be ignored. In the terminology of the financial market, an operator is *long* when buying futures contracts and *short* when selling them. A futures contract obliges the seller (or the buyer) to sell (buy) a specific number of paper barrels at a certain price on an established date. When that day arrives, the transaction is settled in cash, paying the difference between the contract price and the market price in effect at that moment. In a zero-sum game, someone gains and someone loses, but there is no movement of physical petroleum.

There are two regulated financial markets for petroleum, each based on a particular reference crude oil. On the New York Mercantile Exchange (NYMEX), this crude oil is West Texas Intermediate (WTI). The International Commodity Exchange (ICE) in London is based on Brent crude oil.

However, many financial petroleum transactions take place through unregulated bilateral contracts on the over-the-counter (OTC) markets, where the two parties can freely undertake to sell and buy paper barrels reciprocally. Information about transactions on these markets is fragmentary, but according to many estimates, activity on the OTC markets is much higher than on the regulated markets.

The importance of the petroleum futures market has increased because of growing problems with the reliability of prices on the spot market, where physical barrels of petroleum are traded daily. As the production of the traditional benchmarks for the spot market (Brent and WTI) fell—particularly after the early 1990s—indexing the markets to them caused several distortions. The lack of daily availability of these crude oils on the market creates a situation in which a single transaction, for example, a single load of Brent, could be *squeezed* (manipulated), causing its price to shoot up or collapse in a single day. This would send a misleading message to the entire market.

Financial markets seem to offer a more reliable alternative for setting prices, because of several perceived advantages: a much higher number of operators (commercial and financial) with different interests, many more transactions, and much greater liquidity. In other words, these markets appear less subject to price manipulation than the physical crude oil markets.[29]

The impressive growth in the paper barrel market has also been fed by the enormous increase in financial instruments, hedge funds, and financial operators since the 1990s, along with the race to invest in commodities that characterized the first years of the new century. These factors caused the average number of daily positions for petroleum futures on the NYMEX to increase from 272,262 in 1990 to 468,109 in 2000 and 1,303,664 in 2007, when it peaked.[30] Considering that each contract refers to 1,000 barrels, approaching the peak in 2007, more than 1.4 billion paper barrels were exchanged each day just on the NYMEX. This datum captured public attention, because during the same period, world consumption of physical petroleum was less than 85 mbd and production of WTI (the NYMEX reference crude) was less than 300,000 barrels per day!

During the steep rise in prices of 2007–2008, it was easy for many to suspect that this paper barrel explosion was an immense movement of speculation capable of upsetting the trend of real prices. However, this thesis has not proven out yet. Studies conducted on the petroleum financial market and investigations on the subject, including the one launched by the U.S. Senate and assigned to the Commodity Futures Trading Commission (CFTC) did not find evidence that speculation had played a major role in inflating oil prices.

After many years of experience in the petroleum sector, my personal conviction is that although financial operators play a relatively modest role in the petroleum financial market, they can influence its movements by operating on its margins. This will require serious oversight by public authorities, in particular on the role played by some financial institutions, such as investment banks, which could be tainted by conflicts of interest. On one hand, they can publish their analyses and recommendations concerning trends and future prices in the oil market, while on the other, they can trade petroleum futures and other derivatives. In any case, this all has yet to be proven.

More probably, the financial markets can amplify the rise and fall of oil prices when they are already rising or falling. For short periods, the financial markets could push prices contrary to the physical rules of spare capacity, supply, and demand. All this would be almost impossible if the market were perfectly informed and the expectations on the market were based on good data. Unfortunately, in the absence of reliable information, we only know that the financial dimension of petroleum market adds yet another element of uncertainty to the future of the price of oil.

An Ideal World

History has shown that low prices for petroleum or excessive uncertainty about future price movements are the worst enemies of research and development into alternative energy sources. Low prices for crude oil also foster unacceptable consumption habits and discourage energy efficiency. In brief, cheap oil tends to kill sustainable development. A less volatile and more predictable petroleum market would be a reasonable goal, but in practice, it is a chimera.

The price of petroleum is the result of a complex and very delicate equation. The ideal price to foster sustainable development must take into account at least six fundamental elements, some in conflict with others.

Before listing them, I owe you a warning. The assumptions listed below are based on the realities of today; they could change, depending on the evolution of demand, reserves, technology, the value of the U.S. dollar, advances in other energy sources, and other factors.

First, oil prices should be high enough to encourage investment in alternative energy sources, especially renewable energy. With rare exceptions, renewable energy today cannot compete with petroleum when the latter is priced below $50 per barrel. In many cases, it cannot compete when oil prices are below $70 per barrel. To sustain investment in renewable energy over time, crude oil should cost at least $70 per barrel. If we were to penalize carbon dioxide emissions from fossil sources by introducing a tax of, say, $50 per ton of CO_2, this range could be lowered by $10–$15 per barrel.

Second, prices should be low enough not to stunt economic growth. The price of oil above which economic growth suffers is still subject to debate. Nevertheless, various empirical indexes lead me to believe that during the crisis of the early twenty-first century, *demand destruction* began with crude oil prices above $70–$80 per barrel. At least, this was the case in countries like the United States, which neither subsidized the price of petroleum products nor taxed them so much as to distort the final retail price. There remains some doubt that such a range could be sustained for long periods; the high oil prices after

2003 followed two decades of strong economic growth and were also fed by the credit and lending bubble. Considering these elements, a more acceptable range might be $50–$70 per barrel.

Third, prices should be high enough to contain the waste of petroleum and promote energy efficiency. It is not easy to calculate a price that would discourage waste and inefficiency. Some Western residents seem to consider cheap oil and the freedom to waste it to be inalienable rights. Using the United States as a benchmark, it would seem that a price between $70 and $80 per barrel would foster serious conservation. Here again, the range might be lower, say, $50–$70 per barrel, considering the impact of easy credit on consumption habits during the first decade of the century.

Fourth, prices should be low enough to prevent a massive displacement of worldwide agriculture into the production of biofuels. This problem is often underestimated. At the beginning of the century, we witnessed a massive movement to use land for the production of biofuels, which partially contributed to the food crisis of 2007–2008. Probably an oil price above $70 per barrel can trigger such a process, making it more profitable to produce biofuels than food for people or livestock. Such a shift could threaten the poorest part of the world's population.

Fifth, prices should be high enough to assure producers of an adequate return on their investment and to maintain adequate spare capacity. There is much confusion about a fair level for oil prices. I maintain that a long-term price of $50 per barrel should be satisfactory for most players in the sector.

Sixth, prices should be high enough to encourage the implementation of new technologies, both to extract more oil from mature deposits and to drastically abate the environmental and climatic impact of the extraction and use of petroleum. In this case, the ideal price should be no less than $50 per barrel.

This basket of prices seems to center on a price band for the long term (twenty years) between $50 and $60 per barrel. This band seems to be capable of independently sustaining development of our energy and the environmental future of our planet.

Unfortunately, achieving and sustaining an ideal price range is beyond anyone's direct control. As we have seen, the big multinational companies who are the target of much public indignation are themselves second-rate players on the market. OPEC can influence the price of oil indirectly and only briefly by setting production quotas to avoid excess supply to the market. Its theoretical power is severely limited by the undisciplined behavior of its members. In any case, non-OPEC producers, in fierce competition among themselves, control about 60 percent of world production. Thus, OPEC is not an effective cartel, perhaps not a cartel at all.

Popular conspiracy theories notwithstanding, no individual or group can impose and maintain an ideal price range.

A Possible World

Options for containing crude oil prices within an ideal band are few and difficult to realize. Nevertheless, given the importance of doing so, we must make some attempt.

First, we should try to bring more certainty to the petroleum market so that poor data and inadequate analysis do not send mistaken signals and distort reality. This problem could be confronted by a new worldwide energy entity, a sort of Global Energy Agency. Its principal task would be to collect transparent data concerning supply, demand, inventory, production capacity, and so forth for all the countries in the world. This could overcome the principal limitation of the IEA, which was established by the industrialized countries after the 1973 oil shock to safeguard their supplies. The IEA still includes only those countries.

Second, we need to find ways to reward the creation of spare capacity. One possibility might be a Global Stabilization Fund, financed by a very small excise tax paid by consumers to remunerate those producers willing to create new spare capacity. An independent entity like the Global Energy Agency would manage this fund. The agency would be responsible for making the loans and certifying the actual creation of spare capacity.

Another possibility might be the creation of a specific market for spare capacity, financed by the fund described above, in which both producers who create spare capacity and consumers who reduce their own energy consumption could be rewarded.

I realize that these proposals would be very complex to realize. It is certain that without mechanisms for greater transparency and stabilization, the petroleum market will continue to fall prey to the boom-and-bust cycles that have scarred its history.

If we must continue to live with this cyclical scenario, we must try to contain the growth of petroleum demand and noticeably reduce the environmental effects of burning oil. To do this, action is needed on three fronts.

The first is to move the use of petroleum mainly to the transportation sector, where it is virtually irreplaceable, using more environmentally friendly energy sources (gas or nuclear power) or sources with zero environmental impact (sun and wind) to produce electricity, heat, and to meet other nontransportation needs.

The second is to make motor vehicles more efficient, with very low emissions of pollutants and CO_2, through legislative action if necessary. These measures should include regulations that prohibit or seriously restrict the circulation of vehicles with large engines, low mileage, or high speed. For example, we should limit circulation in urban areas to vehicles with very small engines and low/zero emissions. This suggestion could make me the target of many motorists' outrage, but I really see no alternative. It seems farcical that speed limits are imposed throughout the world, while car manufacturers and their customers vie for features that clearly disregard those limits. If the Chinese, the Indians, and the populations in other developing countries were to adopt the consumption patterns of today's Western transportation systems, the effects on the environment and on petroleum consumption would be disastrous. Fortunately, China and other emerging countries are taking this problem seriously, enforcing laws that forbid private large-car circulation in some cities and devising brand-new areas where public transportation will be the only way for people to move.

The third is to drastically abate the emissions and pollutants that oil produces over the course of its use. This will entail a major effort in terms of research and technological innovations. It would be a mistake for universities to continue to devote nearly all their attention to energy sources other than oil, because the search for a rival for crude oil will be long and difficult.

We will be obliged to live with oil for many decades. To make this situation less dangerous for our planet, we must get used to using oil better—by using less of it and making it cleaner. This will be impossible if oil and other fossil fuels remain almost free, as they were for most of the twentieth century. As I will explain in the final chapter, terminating this cheap availability is the first, fundamental step that all responsible governments should undertake to start building a different energy future.

Statistics on Petroleum

Table 1.1. Proven petroleum reserves and reserve life (Top 10 countries and world total, 2008)

Country	*Proven reserves (billions of barrels)*	*Reserve life (years)*
Saudi Arabia	267	68
Iran	136	88
Iraq	115	131
Kuwait	104	103
Venezuela	99	92
United Arab Emirates	98	86
Russia	60	16
Libya	44	61
Nigeria	36	46
Kazakhstan	30	58
World total	**1,178**	**39**

Table 1.2. Petroleum production (Top 10 countries and world total, 2008)

Country	*Production (millions of barrels per day)*
Saudi Arabia	10.4
Russia	10.0
United States	7.5
Iran	4.3
China	3.8
Canada	3.2
Mexico	3.2
United Arab Emirates	3.1
Venezuela	2.6
Kuwait	2.5
World total	**83.8**

Table 1.3. Petroleum consumption (Top 10 countries and world total, 2008)

Country	*Consumption (millions of barrels per day)*
United States	19.9
China	7.9
Japan	4.7
India	3.1
Russia	2.9
Germany	2.6
Brazil	2.5
Saudi Arabia	2.4
Canada	2.3
South Korea	2.2
World total	**85.7**

Table 1.4. Principal uses of petroleum, worldwide by sector (2008)

Sector	*Share (%)*
Electricity	6.7
Heat	0.3
Other energy transformation sectors	6.3
Manufacturing	7.5
Civil and agricultural	11.8
Road transportation	40.4
Other transportation	12.7
Nonenergy uses	14.3

CHAPTER 2

The Splendor and Misery of Coal

I must confess that I have always had a prejudice against coal. This is probably the result of my boyhood reading of desperate, dark stories featuring coal, books by nineteenth- and twentieth-century authors like Charles Dickens, A. J. Cronin (*The Stars Look Down*), or Richard Llewellyn (*How Green Was My Valley*). These authors shaped my feelings about coal as an ugly, dirty source from the past, best forgotten if possible. As I grew older, however, my feelings about coal evolved. I especially forced myself to study it objectively, not to let my judgment fall prey to my prejudice.

No doubt, coal remains the dirtiest and most polluting of all energy sources. It has also caused the greatest loss of human life and produced the greatest devastation on the environment. Yet there are good reasons for its prominent position in the contemporary energy picture. As a matter of fact, coal is the "Prince of Energy." It is second only to oil as a primary source of energy. It is by far the first source for generating electricity, and its consumption grows every year. The story of coal stars not only poor countries and countries with explosive growth like China and India but also the United States, where coal provides half of the electricity.

Its durability over time is due to three factors: cost, availability, and the possibility of making it less dangerous and polluting. Coal is cheap, and the planet has enormous reserves of it. These allowed it to maintain a principal position in the twentieth century in spite of the rapid growth of other energy sources. Recently, however, this preeminence has been challenged by two powerful and converging considerations: environmental protection and climate change. What a bizarre outcome for a raw material that was developed to prevent an environmental disaster!

The modern era of coal began about the middle of the seventeenth century, born of the need to find an alternative energy source to wood. Until then, wood was not only a good material for building houses and ships but also the most commonly used source of heat for working iron and glass. Deforestation became a problem throughout Europe. By the seventeenth century, English iron and glass manufacturers were importing timber from Scotland, Ireland, Wales, and Norway because England had been almost completely deforested.[1] The Age of Coal made the Industrial Revolution possible and transformed the ancient world into the modern world. Without coal, the world would not be what it has become.

Petroleum did not displace coal as the leading energy source until the mid-1960s. Oil was more efficient, cleaner, more flexible, and above all uniquely well suited for transportation. Nuclear power, natural gas, growing environmentalism, and climate concerns seemed to gang up on coal, the resource that had changed the face of preindustrial society. Its share of the world energy mix continued to decrease until 2000. The coal industry and coal-burning utilities reacted with various remedies over time. Technologies for "clean coal" have emerged to reduce its environmental impact and global warming. However, the solutions are not as simple as they sound. "Clean coal" may appear to be an oxymoron, like "cold sun" or "tropical iceberg." Yet cleaner coal is possible. A state-of-the art coal-powered power plant today may achieve a 30 percent higher energy efficiency compared to the older ones, producing less carbon dioxide to obtain the same amount of energy. Furthermore, it may reduce dramatically—in some cases to zero—the emission of other pollutants.

The basic problem is that the entire life cycle of coal is detrimental to the environment and climate, but cleaning it up is difficult because the world has inherited a complex coal system that is too expensive to modify. In fact, the expression "clean coal" today usually refers to burning coal in power plants with reduced carbon dioxide emissions. However, combustion is only the last stage of the uneasy life of coal.

The first phase in the life of coal is its extraction. This phase can devastate entire geographic areas, cutting the tops off mountains and hills or sending people to face deadly danger deep underground.

In particular, this occurs still today in most emerging or transitional countries, where safety is not always a top priority. In China alone, coal mining kills about six thousand people each year. Coal mining areas are desolate places, where the coal dust mixes with water to create sulfuric acid and where the contamination of the groundwater lasts a long time.

The uneasy life of coal continues with its transportation. Because almost 90 percent of the coal burned in the world is consumed by the countries that produce it, coal travels within these countries by rail or by truck, producing additional environmental damage during the trip.

The story of coal ends when it is burned in power stations or furnaces. Even on its deathbed, coal leaves a legacy long past its own demise. When coal is burned, it releases more carbon dioxide than either oil or natural gas does. It should come as no surprise that activities associated with coal represent the largest source of carbon dioxide released by human activity (about 25 percent of the total). Most of the attention of the media and the policy makers is focused on this aspect, but CO_2 is far from being the only burden that coal leaves to our heirs. For example, coal-fired power plants are responsible for half of the more than 2,000 tonnes of mercury dispersed into the atmosphere each year.

As a consequence, to really clean coal, we need to act on it from the moment it is extracted until the moment it is burned. We can never achieve a completely clean coal, but we could at least make it comparable to petroleum (which has its own faults). It will never be as clean as natural gas. Cleaning up the entire life cycle of coal is not technically possible today, and the costs would cancel its economic advantages. That is the big problem with coal.

Emerging countries that have great quantities of it, including China and India, cannot give up on it, or they risk stunting their economic growth. Nor can the most advanced country in the world, the United States, give up on it, because it has the largest reserves and is the world's second largest consumer of coal. Above all, it is highly improbable that the big coal consumers will gladly make the massive investments needed to clean up their consolidated coal systems. This is not because their governments are irresponsible. If we must have coal at low cost, then we must all share the blame. We cannot give up on coal in the age of the Internet, of mass communications, and the global village. Each time we boot up our computers, connect to the Internet, make a phone call, or plug in an appliance, we consume the same energy that illuminates and heats our houses. In many parts of the world, that energy comes from coal.

We may not like this compromise that lets coal prosper, but unfortunately, we are all responsible for it. Still, this does not mean that there is no hope for the future. Coal can become cleaner if we accept that we must pay more for it. Advances in technology may help make cleaner coal cost less in the future, but that future is not around the corner.

What Is Coal?

Coal is organic sedimentary rock, formed from the transformation of vegetable matter over millions of years. The process of forming coal is called *carbonization*. It is really an aging process, which progressively increases the carbon content of the original vegetable remains (in percentage terms) at the expense of hydrogen and oxygen. Thus, coal contains many atoms of carbon and very few atoms of hydrogen, unlike other fossil fuels. This feature of coal is the source of its current problems. In general terms, the higher the amount of hydrogen per unit/weight, the higher the energy density. At the same time, the higher the presence of carbon, the higher the emission of carbon dioxide. Thus, when coal burns, it releases more than twice the carbon dioxide of natural gas and about 50 percent more than petroleum. These percentages also depend on the fact that coal-fired power plants are less efficient than gas-fired ones in particular.

On average, one ton of liquefied gas contains twice as much energy as one ton of coal, and oil contains about 60 percent more than coal.[2] In terms of cost, this is not a problem, because a ton of coal costs a minimal fraction of the same weight of oil or gas. What makes coal more pernicious than the other fossil fuels is its impact on the environment and the climate.

Coal contains inorganic minerals and potentially dangerous components.[3] In 1990, the Clean Air Act identified eleven environmentally toxic metals present in coal. Among these, mercury is the most dangerous. These substances are found in the residues of coal production, as well as in the ash and smoke coming from the combustion of coal. They constitute a serious pollution problem.

However, like petroleum, all coal is not created equal. The term *coal* indicates a family of organic compounds, which may be classified in various ways.[4]

One of the simplest and most widespread classifications of coal is by geologic age, that is, the degree of carbonization of the rock. This method also helps identify the principal energy characteristics of the type of coal being examined. The further along the process of carbonization is (that is, the older the carbonized formation), the higher the percentage content of carbon, the lower the percentage of water and volatile components, and the greater the thermal value. Five stages of carbonization may be identified, but there are only four types of coal.

The first stage involves the formation of *peat*. A mixture of humus and vegetable detritus that has not yet decomposed, peat is widely available and relatively easy to obtain. It covers about 1.5 percent of the surface of the Earth, especially in the Northern Hemisphere. It has a high water content and a low carbon content; therefore, it has a low thermal value. Technically, peat is not classified as coal, but it can be burned to obtain heat.

The next stage of carbonization produces *lignite*, also known as *brown coal*, in which the carbon content can reach just over 35 percent, with up to 50 percent water and a low thermal value of about 3,500–4,600 kilocalories per kilogram (kcal/kg).[5] Dark brown in color, lignite is soft and is used mainly in the *thermoelectric* sector (generating heat and electricity) of poor or developing countries. It is classified as the lowest level of coal.

As time passes, lignite turns into a darker, stronger rock, still with low carbon content (up to 50 percent) but with a higher thermal value (between 4,600 and 6,400 kcal/kg). This is *sub-bituminous coal*, the third stage of carbonization. This rock is used for electrical power generation, for cement making, and throughout industry in general. Lignite and sub-bituminous coal are also known as *soft coals*.

The fourth stage comprises the *bituminous coals*. Their carbon content can reach 86 percent, and their thermal value is more than 6,400 kcal/kg. Black, hard, and with a very low water content, they are the most prevalent form of coal, constituting more than half of the coal consumed in the world. Bituminous coal can be divided into *steam coal* and *coking coal*. The former is widely used in electrical power plants, cement factories, and industry; the latter is used to make steel.[6]

The fifth stage gives us the oldest coal, *anthracite*. Brilliant black in color, it feels like vitrified rock, with a carbon content between 86 percent and 98 percent. Its use is limited, both because of its combustion characteristics (a small flame and an elevated ventilation requirement) and because of its scarcity in nature. Bituminous coal and anthracite are also known as *hard coals*.

Unlike oil or gas, coal is solid. This means that extracting it is like performing crude surgery on the Earth, removing entire parts of the subsoil or even large portions of the surface. The damage caused by the extraction of coal cannot be compared to the extraction of petroleum or gas. After extraction, coal is treated to give it the purity and qualities needed for commercial use. Then it is transported to the consumer market. Transporting coal is much more expensive than extracting it. Transportation can account for 70 percent of its final cost on the market. This is a function of both the difficulty of transporting a solid (coal) compared to a liquid (oil) and the lower energy content of coal. Oil displaced coal as the top energy source partly because oil is ductile and cheaper to transport, and partly because so much more coal has to be moved to obtain the same amount of energy.

The transportation problems are the reason that the large producers of coal are also its principal consumers, while international commerce is rather modest. Overall, only slightly more than 10 percent of the coal produced in the world is exported (for petroleum, the figure is more than 60 percent).[7]

For some time, there have been coal slurry pipelines through which the mineral can be pumped after being pulverized and mixed with water. Yet today, most coal is transported by railroad or by truck, across very long distances with inevitable side effects: Some of the coal is lost on the road, while the long lines of trucks or railcars leave coal dust all along their journey. For example, in China the internal transportation of coal is a critical problem. The large coal-producing centers are very far from the consumption centers, and truck transportation plays a vital role.

There are also technologies for transforming coal into a synthetic fuel, known as *coal liquefaction*. Germany pioneered these technologies, and in the first decade of the twentieth century perfected two chemical processes (Bergius and Fischer-Tropsch) to produce synthetic fuels through *coal hydrogenation* (the reaction of hydrogen and coal).[8] Although these fuels were much more expensive than petroleum derivatives, in 1936 Germany launched a massive plan to build thirty plants to produce synthetic fuels, a project that Adolf Hitler followed personally.

After the war, the Nazi experiments to free themselves from petroleum fed enduring legends, like the one in the film *The Formula*, in which a mysterious Marlon Brando tells how the Germans were able to discover a low-cost synthetic fuel, but that the formula for it was hidden by a worldwide conspiracy led by American multinational petroleum companies. Of course, nothing of the sort ever happened.

The truth was that the coal hydrogenation proved to be too expensive and inefficient, and totally inadequate for solving the Nazis' energy problems. Although many countries pursue it for environmental reasons, coal liquefaction is still restricted by the same limits that made it ineffective for the Germans. Today, liquefied coal accounts for barely 0.6 percent of world consumption, limited essentially to South Africa.[9]

The Resource with Nine Lives

Coal dominated the world energy scene from the eighteenth century until the 1960s, when it yielded to petroleum. It continued to retreat until 2000, when it reached a historical low of a little more than 23 percent of the energy basket of humanity. Since then, however, it has surged again.

Currently, coal accounts for more than one-quarter of total primary energy consumption worldwide. About two-thirds of the demand comes from the thermoelectric sector (heat and electricity), where the dominance of coal is uncontested, generating 40 percent of the electrical power in the world.[10]

Proven reserves of coal are still extremely abundant. At the end of 2008, they totaled more than 910 billion tonnes,[11] almost equally divided between hard coal (53 percent) and soft coal (47 percent). Considering that annual production is about 6.2 billion tonnes,[12] its residual life is a good 147 years.

However, like oil, the proven reserves of coal are only a slice of the resources available worldwide. Like oil as well, the history of coal is full of recurring forecasts of depletion. One of the earliest and most famous was that of William Stanley Jevons, a founder of neoclassical economics. In 1865, Jevons predicted that the coal reserves of Great Britain would be depleted in two decades.[13] Others echoed him, extending their gloomy predictions to the whole world. Yet today the known reserves of coal are far larger than those that existed in the nineteenth and twentieth centuries, after two hundred years of massive consumption. Once again, this teaches us not to trust claims about the amount of resources hidden in the subsoil of the Earth.

Coal reserves are even more concentrated geographically than petroleum. Only six countries share 80 percent of the coal available worldwide. Leading the group is the United States. With more than a quarter of world's coal reserves, it is considered the "Saudi Arabia of coal." The other coal-rich countries are Russia (17.3 percent), China (12.6 percent), India (10.2 percent), Australia (8.6 percent), and South Africa (5.4 percent). These very countries are also the main producers and consumers of coal. Together, they produce 80 percent of the world's coal and consume more than 90 percent of what they produce.

China is the top producer and consumer in the world (about 40 percent of the total). In this country, coal represents 63 percent of primary energy consumption. About half of that is used to generate electricity and heat. These figures show that coal has been (and remains) the lifeblood of the Chinese economic boom.

Between 2002 and 2008, the Chinese demand for coal doubled. In the two-year period 2007–2008, China built coal-fired electric power plants totaling 170 gigawatts (GW). This is a stunning figure. In only two years, China brought on line twice the electrical capacity that Great Britain installed over more than a century. The problem is that this trend has been going on for at least six years, during which it is estimated that around two coal-fired electrical power plants came on line in China each week (each capable of 500 MW). This explains why, as early as 2007, China overtook the United States to capture the unenviable distinction of principal emitter of carbon dioxide.

The Chinese hunger for coal reverberated on world markets, where China alone caused more than 70 percent of the increase in demand from 2003 to 2008. If we add India, the percentage rises to more than 80 percent. Thus, it is easy to understand why attention turns to these two Asian nations when talking about the future of coal and its environmental impact. Chinese and Indian voracity for coal is not likely to abate. By the time the Kyoto Protocol expires in

2012, these two countries will have built almost eight hundred coal-fired electrical power plants, spewing between three and five times as much as much carbon dioxide as the Protocol set as the world-wide target for CO_2 reduction.

However, many developed countries also rely heavily on this fossil fuel with nine lives. Although coal production is rather low except in the United States, coal still represents 20 percent of the primary energy consumption in the industrialized world.[14] What really strikes the imagination, however, is the importance of coal in the United States. Almost a quarter of the primary energy requirement of the country is met by coal. Because American energy consumption is enormous, the demand for this fuel is very high in absolute terms, representing a fifth of the world total. More than 90 percent of the coal used in the United States goes to the thermoelectric sector. Coal produces half of the electrical energy in the country.

This brief outline of the global patterns of coal consumption reveals a crucial fact: Together, the United States and China account for 55 percent of the world's total coal consumption. Thus, these two countries hold the major responsibility of addressing the problem of curbing CO_2 emissions produced by the combustion of coal.

Production costs depend on the origin. They are well contained in developing countries, partly because of the very low cost of labor and the lack of safety and environmental protection. Conversely, they are much higher in the industrialized countries, with the exception of the United States and Australia. They are particularly high in Europe. In 2005, production costs exceeded $120 per tonne in Germany, Spain, and Hungary, while the price of imported coal was about half that. Only government subsidies made it possible for mines in Bulgaria, Germany, Hungary, Romania, and Spain to stay in business and sell their product. As with oil, the price of coal did climb between 2000 and 2008, exceeding $200 per tonne on international markets in 2007 after starting at $35 per tonne in 2000, but that was an anomalous peak. Normally, the international price is less than $70 per tonne. The transportation costs for coal are conspicuous, but so is the investment required for a coal-fired power plant. They are much higher than for a gas-fired electrical power plant (gas being the principal competitor of coal in the electricity sector). The difference in capital costs has grown recently, because coal-fired plants have needed expensive upgrades to meet ever more stringent pollutant emission standards.

Currently, a modern coal-fired plant can cost between two and three times as much as a gas-fired combined-cycle plant of the same electrical generating capacity. Furthermore, coal-fired electrical power stations are less efficient than the new gas-fired ones (56 percent for gas-fired plants, compared to 42–46 percent for a coal-fired plant using the best available technologies, and just 30–35 percent for the better old plants).

Even considering all these elements, coal remains the cheapest energy source for producing electricity. The final price depends on the production cost of coal and the capital cost of building a power plant at a particular point in time. During the first decade of the twenty-first century, the retail price for 1 kilowatt-hour (kWh) from a coal-fired plant in the United States varied between three and five cents. The same kilowatt-hour from natural gas costs between five and eight cents. If we consider the relative price of the three fossil fuels in U.S. dollars per million British thermal unit (MBTU—the unit of measurement commonly used to compare different sources of energy), the comparison is more stunning. For example, in October 2009, I calculated that in the United States, oil (WTI) was priced at $12.5 per MBTU, natural gas at $4.5 per MBTU, and coal at only $2.2 per MBTU.

No other energy source can match these prices, so coal has survived the passing of its golden age, rising from its own ashes.

The Dark Curse of Coal

The principal curse of coal as an energy source is that it pollutes the most and has the worst impact on the climate. Coal is born "dirtier" than other fossil fuels, with an elevated content of carbon, sulfur, mercury, arsenic, other heavy metals, and other substances. These produce great quantities of unhealthful substances when coal is extracted, transported, and burned.

The potential environmental damage of coal begins with its extraction. To understand the extent of this damage, it is enough to take a brief tour through Appalachia, one of the main American coal-producing regions. As summarized by Jeff Goodell:

> In Appalachia alone, the waste from mountaintop removal mining (instead of removing the coal from the mountain, the

> mountain is removed from the coal) has buried more than 1,200 miles of streams, polluted the region's groundwater and rivers, and turned about 400,000 acres of some of the world's most biologically rich temperate forests into flat, barren wastelands.[15]

The situation in Appalachia is only one example of the consequences of *surface mining* (also known as *open cast mining* or *open cut mining*), a system that produces about 60 percent of the coal in the United States. Worldwide, 60 percent of coal is produced by traditional mining, because the coal is deeper underground.[16]

Naturally, mining underground at depths that may range from 500 feet to more than 2,000 feet is more dangerous for the people involved. For example, in China alone, on average six thousand people per year have died producing coal in the last three years. Although the mining death toll has not reached these numbers elsewhere in the world, coal still extracts the highest toll in human life of any energy source.

Underground digging and surface mines involve varying levels of environmental destruction, dust formation, soil erosion, surface water and groundwater pollution, and subsidence. Coal mines also can emit significant amounts of methane, the same natural gas that we burn to produce electricity. The problem is that the methane released during coal extraction is uncontrolled. The risk of explosions is well known. In addition, when it is released unburned, methane has a global warming potential more than twenty times that of carbon dioxide.

Another side effect of coal mining is the formation of sulfuric acid, which occurs when coal waste or dust comes in contact with water. Sulfuric acid is corrosive and caustic. It can pollute the soil and the groundwater, and it contributes to acid rain.

Coal goes on to cause environmental damage during transportation and storage, although there are preventive measures. For example, coal dust can be sprayed with water and the stored coal can be compacted. Adequately covered and protected conveyor belts can keep off rainwater and prevent spreading coal dust between coal car and storage bin and on the way to the furnace. Proper design of coal storage facilities can limit surface runoff water. In addition, contaminated water can be treated before being reused or discharged. All these measures increase the final cost of coal, however, so they are rarely implemented in much of the world.

Finally, there are many problems with the combustion of coal. Compared to its most direct competitor, natural gas, the production of 1 kWh of electricity from coal releases more than twice as much nitrogen oxide, eight times as much heavy metal, and ten times as much fine soot (particulates). Natural gas does not generate sulfur oxides, but coal emits great quantities of them. More than 2,000 tonnes of mercury are dispersed into the atmosphere every year, and it is estimated that half of it comes from burning coal,[17] whereas natural gas does not contain mercury.

Then there are the greenhouse gases. As we have seen, even the most advanced coal-fired plants emit twice as much CO_2 as natural gas to produce 1 kWh of electrical energy. The data on carbon dioxide emissions show the impact of burning this fossil fuel and the sectors and countries most involved.[18] Each year, coal produces almost 12 billion tonnes of CO_2, about one-quarter of the worldwide emissions of anthropogenic (human-caused) CO_2. The thermoelectric sector accounts for two-thirds of this.[19]

Is "Clean Coal" Possible?

Even the fiercest opponents of coal must concede that tremendous technical progress has been achieved in coal conversion to heat and electricity in the last thirty years, in terms of both efficiency and the reduction of pollutant emissions, particularly of nitrogen oxides (NO_x), sulfur dioxide (SO_2), and particulates. The results of these efforts can be grouped into three categories of "clean coal technologies:"[20]

1. *Supercritical and Ultra-Supercritical pulverized coal combustion.* This involves burning very fine coal dust at very high pressures and temperatures.[21] This noticeably improves the efficiency of the production cycle, reaching 42–46 percent, compared to 30–35 percent for a traditional plant. Ultra-supercritical combustion also reduces the amount of pollution for a comparable amount of electrical energy generated. If we add low-emission burners and more advanced denitrifiers, desulfurizers, and smoke filters, we can achieve a reduction in nitrogen oxides, sulfur dioxides, and particulates of between 90 percent and 100 percent. Obviously, the more reduction there is, the higher the cost.

2. *Fluidized-bed combustion*. Pulverized coal is injected into a fluid bed of heated particles (a mixture of salts, mainly lime). The coal is suspended first by a jet of air and then by its own combustion gases. The salt mixture absorbs the ash and some of the harmful gases from the burning coal, drastically reducing emissions of sulfur oxides. It also operates at lower combustion temperatures, which reduces the formation of nitrogen oxides. These plants also need smoke filters to obtain high levels of pollution abatement. Their advantage is that they can burn a wide range of fuels, even biomass (wood).
3. *Integrated gasification combined cycle* (IGCC). In this case, pulverized coal is brought into contact with steam and oxygen at high pressure and temperature. This causes chemical reactions that produce synthetic gas (syngas), which consists mostly of hydrogen and carbon oxides. After being cleaned to remove the sulfur, the syngas is sent to a turbine. The hot gas from the turbine discharge generates steam to drive another turbine (similar to the gas-fired combined cycle). Currently, the yield from this technology is close to 40 percent. The sulfur in the coal can be completely recovered in a commercial form, and the ash is converted into inert vitrified slag.

On paper, the big advantage of gasification is that it is easy to integrate with the capture and geological storage of carbon dioxide (see chapter 4). The gas produced from distilled coal has a high concentration of carbon dioxide at high pressure. These are ideal conditions for capturing and storing the CO_2 without using much energy. In the trade, this is called *CO_2 sequestration.*

Unfortunately, these technologies still have limits. The integration of IGCC plants and CO_2 sequestration remains unrealized. Most important, coal-cleaning technologies still cost too much for coal to compete with natural gas. Current estimates indicate that costs per installed kilowatt-hour vary between $1,900 and $2,400 for an ultra-supercritical pulverized coal power plant, and $2,400 and $3,000 for an IGCC plant.[22] By way of comparison, the cost per installed kilowatt-hour of a gas-fired combined-cycle plant varies between $600 and $800, providing better environmental performance and producing less carbon dioxide.

The real issue is that, on average, the world's coal-fired power plants are quite old. In the United States, for instance, the average

age of coal-fired plants is 35 years, and only 10 percent are less than 20 years old. The most efficient and advanced plants, those with supercritical combustion capacity, represent only 23 percent of the total (or 75 GW). The figure is 13 percent worldwide. By contrast, Japan has the most efficient and cleanest inventory of coal-fired plants in the world, with 70 percent of them being supercritical. The age and low efficiency of U.S. power plants (32 percent on average) allow them to produce cheap electricity and sizeable profit margins for their owners. Unfortunately, as in many other parts of the world, this severely complicates efforts to mitigate climate change. A study by the Massachusetts Institute of Technology (MIT) reported, "Shutting down fully amortized plants and replacing them with higher-cost new plants with low CO_2 emissions could have significant economic impact and would likely raise substantial political opposition."[23]

However, widespread adoption of the three "clean coal" technologies would abate pollutants like nitrogen and sulfur oxides, and particulates, but only make a minor dent in the emissions of carbon dioxide that are typical of the combustion of coal. The only way to contain or nullify the CO_2 emissions from coal is to capture the CO_2 and store it. This remains the great hope for the future, even if currently its costs are still too high. So far, there have been too many false starts on this subject. As journalist Kent Garber noted: "In fact, there have been a number of plans to do so, but they haven't turned out so well. In most cases, they have been heralded by news releases and applause. Then, they quietly fell apart."[24] True, there are still no operational plants in the world in which the combustion of coal is connected with the capture and storage of the carbon dioxide. Its problems are so complex that I have devoted a specific chapter to it (see chapter 4).

Cleaner coal is thus possible, but it would require a revolution in the ways we produce, transport, and use this resource.

The Future

Faced with the daunting problems of starting a coal revolution, many governments put their heads in the sand to avoid dealing with such a plentiful and (apparently) inexpensive source of energy. This situation gives the dark prince of energy another of its nine lives.

Areas like Europe, where there is more awareness of environmental and climatic conditions, will be able to limit their use of coal, largely because they do not possess vast reserves on their territory. Those countries holding immense reserves of low-cost coal will find it very hard to stop in spite of the risks to the environment and climate. Almost 90 percent of the coal consumed in the world is burned in the countries that possess most of it: China, India, and the United States. The situation is even more complex for the two Asian nations. They have plenty of coal, but lack other sources of energy. They must import oil and natural gas, which cost more than homegrown coal.

For this reason, over the next two decades most of the increase in coal consumption will come from China, India, and the emerging countries. Consequently, most of the global increase in CO_2 emissions will come from Asia. In this scenario, the great quantities of greenhouse gases released into the atmosphere by the two Asian nations would completely undo the objectives that Europe and other industrialized countries have set for CO_2 reduction over the next two decades.

The only way to stop this race is for nations such as China and India to accept carbon dioxide ceilings. Although the Kyoto Protocol (1997) provides for ceilings, they have never bound developing nations. Furthermore, the United States has never accepted the Kyoto Protocol. Indeed, after initially promoting it, the U.S. Congress did not ratify the treaty.

So far, the European Union has led the way in imposing firm ceilings on carbon dioxide emissions, initiating a "cap-and-trade" system. Under such a system, each European country distributes CO_2 emission ceilings to its manufacturing sectors. Each manufacturing unit that emits more than its allowed CO_2 volume must pay heavy fines or purchase *emission certificates* on the market. These certificates are sold by those who earn *emission credits* by reducing their own emissions below what they are allowed. The price of the certificates is set by a free market. In theory, if the emission ceilings are calibrated carefully and if CO_2 emissions exceed them substantially, the price for the emission certificates should rise, making it ever more expensive to release this gas into the atmosphere.

Regulation was relatively ineffective during the first three-year period of this scheme (2005–2007) because the European Union had issued too many free emission allowances. After this period,

European emissions regulation was changed, launching a much more rigid regulatory plan for the five-year period 2008–2012. The overall objective is to reduce CO_2 emissions in 2020 by 20 percent compared to 1990. Today it is estimated that these plans should cause the cost of the certificates to rise noticeably, oscillating between €30 and €50 per tonne of CO_2 during the decade 2011–2020.

After the European experience, the United States started working on a similar cap-and-trade model for carbon dioxide. It was a central issue for the Obama campaign in the 2008 elections, and it is part of the Waxman-Markey energy bill working its way through the Congress.

Although criticized from the right and from the left for opposing reasons, the American cap-and-trade system has its merits if we look at the future. Above all, it might finally convince countries like China and India to adopt similar systems for limiting CO_2 emissions. In fact, without involving these two countries in similar mechanisms, any Western effort to cope with the CO_2 problem would be like trying to empty a lake with a bucket.

The flood of CO_2 from Asia is not the only problem. If only the West adopts CO_2 reduction systems, many Western companies might relocate their manufacturing to countries that do not adopt such systems to avoid paying the cost. This phenomenon is called *carbon leakage*. It would undo the best intentions of Western governments and simply shift the CO_2 emissions to another part of the world.

The Obama plan does take a competitive advantage away from the United States, imposing a real cost for emitting CO_2. However, this is what it may take to bring the Chinese and the Indians around. It certainly irks them that the most polluting country in the world reserves the right to pollute without restrictions.

It is easy to overlook the perspective of those developing countries, to whom the environmental concerns of the United States and Europe smack of hypocrisy. At the dawn of the Western Industrial Revolution, only expansion and profit counted. For 250 years, the West revolutionized its economic and manufacturing systems, polluting the environment and warming the atmosphere, while accruing an economic advantage that today it wants to deny to developing countries. If the Obama administration can rebut this argument and lead China and India to participate in a global effort to reduce CO_2 emissions, it will have achieved a historic milestone for the planet and for international relations.

Even without movement on the greenhouse gas front, we must hope that the major coal-using countries will want to do something about local pollution. Those who produce carbon dioxide do not feel it, because it blows away, but local pollution has an immediate impact on the quality of life at the source of the problem.

Although the big producers and consumers of coal did not sign the Kyoto Protocol, they still have to deal with the growing mess at home, the devastating environmental impact of coal and other polluting sources. It is worth noting that China also burns more wood and dried manure for energy than any other country.

Currently, only 6 percent of the coal-fired power stations in China are equipped with sulfur dioxide scrubbers, and we are not even sure that they are always operating.[25] The local level of air pollution has already reached alarming levels, especially in cities near the large coal-fired power plants. The smog that created problems during the construction of the facilities in Beijing for the 2008 Olympics was only a small symptom of the ills across China and other countries.

The other side of unsustainable consumption is waste. In China, India, and most developing countries, energy efficiency is very low. They use more energy than the industrialized countries to obtain the same results. In economic terms, it takes much more energy to obtain one dollar of gross domestic product (GDP). Estimates concerning this phenomenon are uncertain, because transparent, complete data are not made available by the governments involved. Based on what we do have, most estimates indicate that a country like China uses five to ten times as much energy to obtain one dollar of GDP as Japan does.

Many Chinese and Indian power plants were built with outdated technology. They are in a bad state of repair, so that electrical blackouts are a constant feature of the Chinese energy scene. To be fair, this situation is reflected in many countries, including the United States. It is behind the frequent unexpected peaks in Chinese demand for petroleum. When blackouts proliferate, oil is the only substitute that can quickly compensate for the nonperforming coal-fired electrical system. An example was the spike in crude oil demand in China during 2003 and 2004.

Aware of these problems, the Chinese government is subsidizing cutting-edge technical solutions, although they will not yield results soon. These range from coal liquefaction at the mine, allowing transportation by pipeline, to the adoption of clean coal technologies. Furthermore, most of the new coal-fired power plants built by China after 2000 are more efficient that those existing in the United States. In an encouraging development, during the G-20 summit in Pittsburg in September 2009, Chinese President Hu Janthao opened for the first time the possibility of China joining a post-Kyoto system of CO_2 regulation.

Forging a new international agreement to fight climate change after the Kyoto Protocol expires in 2012 will be difficult. The Copenhagen Summit on Climate Change in December 2009 showed that. However, the determination of the Obama administration to make fighting climate change a central point of American politics, and the openness of China to consider a post-Kyoto system of CO_2 emissions mitigation nourish the hope that a new agreement on climate change may be possible before 2012.

The point remains that research and technological innovation are essential to realistically facing the problems of the most damaging fossil fuel. Because part of the damage caused by the combustion of coal and other fossil fuels transcends national borders and becomes a global disaster, the support and financing of scientific research must also be subject to binding agreements on international cooperation.

You may or may not like coal, but a world that lives on electricity cannot do without it. Like petroleum, learning to use coal less and better is one of the great challenges of the twenty-first century. We cannot achieve a paradigm shift in a few years, because neither existing technologies nor people's behavior are mature enough to support a real revolution of the way the world consumes coal. We can hope to achieve this in the future only if we are willing to pay more for coal for the sake of our planet's health. That is what it will take to make coal cleaner and to sustain the research and technological innovation to make it even cleaner. How to deal with the higher electricity and heating costs for consumers is something that I will cover in the general framework suggested in the Conclusions.

But the starting point of that framework is that cheap energy and the fight against climate change are incompatible.

Statistics on Coal

Table 2.1. Proven coal reserves and reserve life (Top 10 countries and world total, 2008)

Country	*Proven reserves (millions of tonnes)**	*Reserve life (years)*
United States	238,308	221
Russian Federation	157,010	480
China	114,500	43
Australia	76,200	190
India	58,600	114
Ukraine	33,873	565
Kazakhstan	31,300	296
South Africa	30,408	121
Poland	7,502	52
Brazil	7,059	1,188
World total	**826,001**	**123**

* Physical tonnes of raw coal (not tonnes of coal equivalent)

Table 2.2. Coal production (Top 10 countries and world total, 2008)

Country	*Production (millions of tonnes)**
China	2,650
United States	1,076
India	514
Australia	401
Russia	327
Indonesia	279
South Africa	251
Germany	195
Poland	146
Kazakhstan	106
World total	**6,724**

* Physical tonnes of raw coal (not tonnes of coal equivalent)

Table 2.3. Coal consumption (Top 10 countries and world total, 2008)

Country	*Consumption (millions of tonnes)**
China	2,671
United States	1,015
India	579
Germany	238
Russia	233
Japan	193
South Africa	192
Australia	144
Poland	141
Turkey	109
World total	**6,669**

* Physical tonnes of raw coal (not tonnes of coal equivalent)

Table 2.4. Principal uses of coal, worldwide by sector (2008)

Sector	*Share (%)*
Electricity	66.4
Heat	2.9
Other energy transformation sectors	2.7
Metallurgy	16.5
Chemicals	1.6
Other manufacturing	4.9
Civil, commercial, and agricultural sectors	3.9
Transportation	0.1
Nonenergy uses	1.0

CHAPTER 3

All That Gas

Natural gas was the poor cousin of petroleum for most of its history, a curse to the drillers who found it, tempered by the fact that the wells also yielded oil. In remote areas, it was burned at the wellhead or vented to the atmosphere, because building a gas pipeline to send it anywhere cost too much. Since the 1980s, things have changed dramatically. Today gas holds third place in world energy consumption, but is also the fossil fuel with the rosiest future and the best prospects for growth. This is because of its molecular structure.

Its principal and most popular component, methane, consists of one atom of carbon and four atoms of hydrogen (CH_4), so it is the simplest fossil fuel molecule. It is also the one with the least amount of carbon, which makes it the cleanest fossil fuel. For the same amount of electrical power generated, burning natural gas emits half the carbon dioxide of coal and about two-thirds that of petroleum products. Plants that use natural gas (especially electrical power plants) are also smaller, easier to build, more flexible, more efficient, and more economical than plants using other fossil fuels.

In the last twenty years, natural gas has spread most in the electrical sector, while still growing in the manufacturing or residential sectors (for heating). Its importance has also increased noticeably in the worldwide petrochemical industry (it was already widely used in the United States). Meanwhile, it remains the principal source of hydrogen at a relatively acceptable cost. The use of natural gas is growing in other areas as well. In gaseous or liquid form, it can be used for transportation, competing with petroleum in the very sector where oil is largely uncontested. In many industrialized countries, methane plants may soon replace aging nuclear power plants that are forced to close or coal-fired plants whose pollution can no longer be tolerated.

Although all these elements seem to augur spectacular success for gas, several problems weigh on its future. To start with, its price: its success is the fruit not only of its environmental advantages but also low cost brought about by decades of great abundance and the relative powerlessness of the producer countries. This picture may evolve in the future. Like prices for petroleum, the price of gas has tripled or quadrupled since the beginning of the twenty-first century, turning the poor cousin of crude oil into a precious resource. Then natural gas prices plummeted after 2008 because of both the recession and a relative overabundance of gas, spurred by investments to develop gas deposits and infrastructure when prices were high. Trends for gas prices, supply, and demand in the future may be more volatile than they were through the end of the last century. Liberalized markets, such as the United States and Great Britain, have already experienced volatility in the first decade of the twentieth-first century, but it is a new phenomenon for Europe.

A more volatile natural gas scenario could lead to a paradigm shift for the European gas market. Volatility could also be relevant for the global market as a whole, depending on the effective rate of development of liquefied natural gas (LNG). In fact, today one of the major limitations to the expansion of natural gas is still its rigid market structure: Only 30 percent of gas produced worldwide is exported (compared to more than 60 percent for oil), and less than one-third of the exported gas goes through vessels in liquid form. Most natural gas moves through gas pipelines that directly link producers and consumers. This rigidity has impeded the creation of a global gas market, which may develop only if a more flexible transportation system develops (like LNG). For now, gas exchanges remain essentially confined to three big regional markets: North America, Europe, and Asia.

Further, resource nationalism and efforts to control both production and prices have spread in the natural gas sector among several big producers. In some cases, the gas producers have more leverage than oil producers because the consumers linked to them by pipelines have no alternative source of supply. This is particularly true for the European Union, which heavily depends on natural gas imports via pipelines from a small number of countries. Thus, there is the possibility that major gas producers (particularly exporters to Europe) may form a gas cartel in the future—a sort of OPEC of natural gas—to achieve their objectives.

Yet natural gas is beginning to present a dilemma even for the producer countries, which often have an increasing need for methane. In the Middle East and North Africa, two areas that produce great quantities of gas, electrical consumption is increasing at a dizzying rate, imposing a greater need for domestic use of gas. Then, for many countries, gas has more value when it is injected into crude oil deposits to increase petroleum production. This makes the injected gas unavailable to international markets.

Another big issue has complicated the future of natural gas even more. The high prices of the first decade of this century made it feasible to apply oil-sector technologies to gas extraction. This has led to a surprising, massive development of unconventional gas resources in the United States, which until a few years ago appeared doomed to an unstoppable decline of its gas production. While this element has changed the perspective of LNG global trading, it has also opened up a new horizon for gas development worldwide.

Finally, the depressed prices of 2008–2009 proved that natural gas is a dreadful competitor of most renewables and a staunch ally of any strategy of carbon dioxide mitigation. Of course, any competition between gas and renewables will depend on how natural gas prices evolve in the near future. How these challenges unfold will shape the future of natural gas.

What Is Natural Gas?

The expression *natural gas* indicates a mixture of fossil fuels, mainly methane, which constitutes on average between 70 percent and 90 percent of it. Once the gas is purified of other components, methane is the gas that we use to heat our homes, produce electrical power, drive our factories, and cook.

Although highly flammable, methane is neither toxic nor poisonous (so it is not a painless way to commit suicide). It is invisible and odorless; the distribution companies use special additives to give methane its familiar smell as a safety measure.

Natural gas at the wellhead contains other hydrocarbons in addition to methane, including propane, butane, and pentane. Each has its own usefulness and popularity. Known popularly as *liquefied petroleum gas* (LPG, not to be confused with LNG), propane and butane have many common uses, for instance, for motor vehicles, cigarette lighters, and backyard gas grills.[1] Unlike methane and ethane, once propane and butane are extracted from the deposits in a gaseous form, they can be liquefied at room temperature by subjecting them to moderate pressure. Thus, they can be easily stored in transportable containers such as gas bottles. Ethane, propane, and butane are also used as raw materials in the petrochemical industry, being the precursors of some of the *building blocks* (basic starting compounds) for synthesizing many petrochemical products.

All natural gas is not the same, and experts in the field have some curious names for the differences. When natural gas is associated with liquid hydrocarbons or can be easily condensed, it is called *wet gas*; otherwise, it is *dry gas*. Gas in deposits that contain petroleum is *associated gas*. Crude gas containing large quantities of sulfur in the form of hydrosulfuric acid is *sour gas*.

Extracted natural gas must be treated before it can be sent to our houses. Even after treatment, methane has different heat values, depending on the deposit from which it comes. Its specific characteristics affect its value (and price). As we saw in the chapter on petroleum, natural gas is often found in the subsoil along with crude oil, because in many cases their genesis was the same. Natural gas is also found in other formations where crude oil could not take form or endure. For example, gas can be found in the subsoil at far greater depths than petroleum, which decomposes under extreme conditions of temperature and pressure.

So far, the world has mainly developed and exploited deposits of *conventional gas*, gas that is economically exploitable using available technology. The relatively new field of unconventional gas is attracting much interest. Broadly, unconventional gas encompasses at least four main categories.

The first is *coal bed methane*, the gas found in coal deposits that can only be extracted together with the water found inside the deposit itself. In fact, the water surrounding coal beds absorbs and retains the methane as long as the coal beds are under pressure. Treating and disposing of the highly concentrated salt water is one of the most significant limits to exploiting coal bed methane. The first case of coal bed methane extracted for commercial use was the Cedar Hill Field development in New Mexico in the 1970s. Before then, natural gas in coal deposits was vented to the atmosphere to reduce the risk of explosions. Several projects to exploit this type of reserve are being studied, especially in Australia and China. On a world level, coal bed methane reserves are estimated to be as large as current proven reserves of conventional natural gas.

The second category is *tight gas*, the gas found in very compact sediments (sandy concrete stones), which can only be freed by stimulating the deposit. Tight gas was the first type of unconventional gas to deliver significant production volumes.

The third is *shale gas*, gas that has accumulated in very porous gas-rich geological structures (typically clays), which, like tight gas deposits, do not allow the gases to flow to the surface when the deposits are perforated. Shale gas probably has highest potential for rapid growth. Because these deposits lie in very thin strata, commercial-scale shale gas extraction can use technologies that originated in the oil sector. These include *horizontal drilling* (coming in from the side) and *hydraulic fracturing*, in which water and sand are injected at high pressure (3,500 pounds per square inch) into wells to shatter rocks deep underground, helping to release the trapped gas.

Although the first production of shale gas goes back to the 1800s, it has captured industry attention only since 2000, thanks to the extraordinary success of the Barnett Shale deposit in Texas. Except for the United States, there are no accurate estimates of global shale gas and tight gas reserves because there has been little exploration. Nevertheless, it is commonly believed that they greatly exceed proven conventional natural gas reserves.

Finally, there are the *methane hydrates,* natural gases trapped inside the crystalline reticules of glacial deposits (mainly in the areas of permafrost, such as Siberia, Alaska, and Canada) or on the bottom of the ocean. Exploration has been very limited, so there are no agreed estimates on the size of methane hydrate deposits. They may be gigantic, on the order of tens or hundreds of times greater than the deposits of conventional natural gas. For now, these resources remain essentially untouched because of the significant technical, cost-related, and environmental problems associated with exploiting them. Even if new technologies allow us to exploit them economically, methane hydrates will likely become only a modest fraction of the total resource base, at least for the next two or three decades.

Unlike petroleum, there is methane in the lithosphere of the Earth and in the solar system. Traces of natural gas have been found on Mars, and in 2004, a European space mission discovered traces on Titan, one of the moons of Jupiter. This has led some scientists to suggest that natural gas may also have an inorganic origin. Initially hypothesized by the Soviet school of petroleum geologists, this idea was strongly rebuffed by Western scientists at first. Lately, it has found some vigorous support in the West,[2] but it has not been accepted by the general scientific community yet.

Here on Earth, total reserves of natural gas available are immense, and even the proven reserves of conventional natural gas, a mere fraction of the others, are more than abundant: about 180,000 billion cubic meters (bcm), compared to annual worldwide consumption of less than 3,000 billion. This means that the residual life of just the proven reserves of gas is more than sixty years, a quarter-century longer than that of petroleum reserves.[3] Notwithstanding these figures, there is no shortage of gas-pessimists (who usually also support Peak Oil theories) who believe that natural gas reserves are overestimated. In their opinion, gas production will also peak in the next few decades. My opinion is the opposite. Natural gas production has even more room than oil to expand in this century, dependent only on cost and technology, as proved by the boom of U.S. shale gas production in the last few years.

As with petroleum and coal, natural gas proven reserves are very concentrated. Only three countries control more than 55 percent of them: Russia, Iran, and Qatar. Almost 80 percent of the conventional gas that can be produced in the coming decades is located in Russia and the former Soviet republics, the Middle East, North Africa, and Nigeria. In spite of its great availability and its environmental advantages, natural gas has had a slower and more complicated industrial evolution than its fossil cousins. Its misfortune is being a gas, which makes it difficult and expensive to move.

It takes about a thousand times more methane by volume than petroleum to provide the same energy content. Oil is easily transported at low cost (by pipeline or by ship) from wherever it is produced to almost anyplace in the world, but for much of the twentieth century this was simply too expensive for gas. Not by chance then, the first gas markets developed near the deposits that could feed them directly through relatively short gas pipelines. Even this option encountered many obstacles.

To be transported in a pipe, methane must be compressed and then introduced into gas pipelines several meters underground (thousands of meters below, to traverse mountains or the ocean bottom). Throughout the trip, the gas pressure must be maintained by suitable compressor units. To avoid the risk of accidents, the welds and the piping materials must be able to resist pressure and corrosion. Control systems along the gas pipeline must monitor the normal movement of the gas and detect any small imperfection in or damage to the structures. This complex and delicate structure makes the transportation cost especially high. On average, laying a gas pipeline onshore capable of transporting 8 bcm per year for less than 1,000 kilometers could require an investment of about $1 to $2 million per kilometer (much depends on the cost of raw materials, especially steel). Complex geographic conditions can triple the expense. If the pipeline has to go offshore, the cost may increase even more. On the other hand, a gas pipeline achieves economies of scale by moving large amounts continuously, diminishing the cost per cubic meter transported.

For much of its history, the production of natural gas has cost much less than its transportation and distribution, which can often represent 80 percent of its industrial cost. The situation is better today, but the transportation and logistics of moving gas over long distances (generally more than 600 miles [1,000 kilometers]) still constitute much of its final cost, often about 50 percent.

Liquefied natural gas has been well known since the early twentieth century. Liquefaction involves cooling the methane to –161°C, causing it to go into a liquid state and reducing its volume by about six hundred times. Liquefaction plants must be located near the sources of extraction, and they require a very high investment because of the safety conditions under which they must operate. Once liquefied, the gas is transported to market in special LNG ships, which keep the methane liquid until final docking. There it is returned to a gaseous state at a regasification terminal and connected to gas pipelines that transport it to the final consumers.

Liquefying natural gas can constitute 70–80 percent of the industrial cost of the final product on the market. This is similar to what happens with coal. There is just no comparison with the ease and economy of moving petroleum, for which transportation affects less than 10 percent of its industrial cost. This complex infrastructure makes the LNG chain cost more than $3 billion for a plant with an annual capacity of 8 bcm in the best-case scenario.[4] As a rule of thumb, the production and logistical costs of the LNG chain (liquefaction, transport, and regasification) are at least twice as much as the costs of transporting natural gas via pipeline (using the same volumes of gas produced and transported as a reference). The LNG chain also uses between 10 percent and 24 percent of its own gas internally in processing and handling (depending on the size of the installation), compared to the internal consumption of gas pipelines, which is usually less than 5 percent. Another significant feature of the LNG picture is a global mismatch in the chain: worldwide liquefaction capacity is around half of the regasification capacity. This implies that, on average, regasification plants work at half of their capacity, and the LNG market is not "liquid" enough to support significant volumes of arbitrage.

Because of the high cost of transportation, more than 70 percent of methane consumption continues to occur in the countries that produce it (only 25 percent in the case of petroleum). Given the relatively high cost of LNG, about three-quarters of what is exported still goes by gas pipelines. For the same reasons, even today it is not cost-effective to develop many small and medium-size gas deposits if they are too far from the consuming markets. Associated gas from oil production is often burned (*gas flaring*) or released into the atmosphere (*gas venting*), with serious environmental consequences.[5] Each year, at least 150 bcm of gas is wasted this way, almost twice the annual consumption of a big methane consumer such as Italy. The full scale of the problem is unknown, because many countries do not report, or underreport, this kind of gas disposal.

Storing gas also has its problems. Methane can be accumulated in exhausted deposits or in underground water or salt caverns with impermeable walls, which prevent the stored gas from escaping. The first problem is the actual availability of suitable deposits or geological formations. The second is maintaining the stored gas under pressure, so that it can be extracted when needed. Some of the gas in the subterranean deposit is needed simply to push out the extracted gas. Gas used for this is called *cushion gas*, while the gas that is extracted is called *working gas*. The cushion gas can be partially extracted when the storage site is abandoned.

Even today, the intrinsic difficulties of transporting and storing natural gas impose a general rigidity on the gas industry and constitute a major obstacle to the creation of a global market for this resource.

Damned Gas! A Brief History of a Difficult Resource

Natural gas was the last fossil fuel to take hold at the world level, although its industrial history is only slightly shorter than that of petroleum. In terms of production and market, that history was essentially limited to the United States for the first seventy years.

The history of gas opened during the era of John D. Rockefeller, the father of the modern petroleum industry, in the late nineteenth century. Pittsburgh was the first city (1883) to receive methane by gas pipeline, but a transportation network was already developing in Pennsylvania, Ohio, and New York. Rockefeller himself set up a trust for natural gas in 1886.

Over the next decade, the American gas industry grew more slowly than the oil industry because the costs and technological problems of transportation made it cost-effective to build gas pipelines only where the deposits were near the potential centers of consumption. The first long-distance gas pipelines were not built until 1927. The first was 250 miles (400 kilometers) long; the second more than 350 miles (560 kilometers). The most important step forward occurred in 1930, when the Natural Gas Pipeline Company of America built a gas pipeline with a capacity of 2 bcm per year from Texas to Chicago, almost 1,000 miles (1,600 kilometers) long, at a cost of $35 million.

American consumers were not easily convinced to use natural gas, given the competitiveness of alternative fuels, mainly coal. For this reason, by the end of the 1930s, almost two-thirds of the gas produced in the United States was vented or flared. Finding gas remained more or less a curse for petroleum hunters, a harbinger of economic failure or devastating accidents. With a market that was too small to absorb increased gas production, as well as stiff competition among companies, the Great Depression dealt another hard blow to the new industry.[6]

During and immediately after World War II, gas began to win a more important role among energy sources, but only in the United States. Fearing that American petroleum was rapidly declining, the administration of President Franklin D. Roosevelt pushed to develop the vast reserves of gas in the western and southwestern parts of the country.[7] The Cold War added the fear of having to fight the Soviet Union openly without the energy self-sufficiency that the United States had always enjoyed. Soon these fears proved to be clearly unfounded, but natural gas benefited from them.

In 1947, two large oil pipelines running from east Texas to the northeastern United States called Little Inch and Big Inch, were converted into gas pipelines, ushering in a new era for the development of methane. By 1950, the United States was already consuming 170 bcm of gas, more than 90 percent of the gas used throughout the world. The Soviet Union was using only 6 bcm; the rest of the world, 11 bcm. From 1950 to 1970, American consumption of natural gas grew at a prodigious rate of 7 percent per year, more than tripling in twenty years, driven by a market regulatory system that kept prices artificially low.[8]

Only at the beginning of the 1960s did natural gas begin to make some room for itself on an industrial scale in other areas of the planet. The Russian Republic of the Soviet Union held the largest gas resources in the world. Consumption in the Soviet Union took off in real terms only at the end of the 1950s, stimulated by a new will to modernize the economy.[9] Europe joined the gas world much later. Throughout the 1950s, natural gas covered only about 1 percent of the primary energy needs of the continent. It was only with the discovery of the massive deposit at Groningen, Netherlands, in 1959 that methane began to break into the European market. The only exception in Europe was Italy, thanks to the Eni of Enrico Mattei, which upgraded the gas reserves of the Po Valley, developing an industry that by 1960 made Italy the top gas consumer in Western Europe and number 5 in the world (after the United States, the Soviet Union, Canada, and Romania).

During the 1960s, there was enormous growth in the demand for gas in the Soviet Union, Europe, and parts of Asia, mainly for heating. The oil shocks of the 1970s raised worldwide awareness of gas as an alternative to petroleum. With the growth of consumption, there came the need to internationalize the market, at least in Europe.

While North America and the Soviet Union were self-sufficient, many European countries had limited reserves of gas and needed to import it, so they built large cross-border gas pipelines from the

Netherlands, the Soviet Union, and later Algeria (1980s). At the time, the Russian and Algerian projects seemed an economic folly and an engineering hazard. In the case of Russian gas, they required connecting the distant gas deposits in western Siberia with the principal consumption centers of Europe. In the case of Algeria, the export of large volumes of gas required building a pipeline across the Mediterranean at record depths to the island of Sicily. However, the risk turned into an extraordinary success. By 1980, Russian gas exports to Western Europe, which had begun during the early 1970s, reached 55 bcm of gas per year. The first Algerian gas to Italy through the Transmed pipeline started flowing in 1983.

As they built up the gas transportation infrastructure, producer and consumer countries also defined the standard contractual agreements that still serve today as the reference for most of the world outside the United States, Canada, and Great Britain, particularly in some of their more peculiar aspects. Contracts for twenty to thirty years featured a unique rule, the *take-or-pay* clause, which required the buyer to pay for a certain volume of gas each year (*minimum annual quantity*), even if drawdowns were less than that or even none at all.

The reason for the take-or-pay clause is simple. While those who produce petroleum and transport it by ship can sell it to any country in the world, a gas producer selling through a pipeline has a limited number of potential customers, namely, those countries crossed by the gas pipeline itself. Given the enormous cost of the transportation infrastructure, the producer needs safe revenues such as those guaranteed by take-or-pay clauses to ensure a sustained return on investment. For their part, the customers accept long-term commitments because they also need to build gas pipelines in their countries and need to be able to count on a reliable supply of gas for many years to come.

This market structure favored the establishment of big European gas companies, later nicknamed "national champions," acting as de facto monopolies in their own countries. They bought the bulk of the gas needed by their domestic markets and took care of building up national infrastructure networks and securing a stable supply. The fact that they were the only buyers increased their leverage vis-à-vis the powerful exporting countries.

Long-term supply commitments proved to be effective instruments for ensuring safe and continued gas supplies for many decades. Only after 2006, particularly after the partial interruption of gas exports to Europe during the Russian-Ukrainian crisis, have they begun to be questioned.

During the 1970s, producers and consumers defined another contractual element that has endured to our times: the way that gas prices are set. To make this competitive, they devised a formula that tied the price of gas to price trends for the principal competing fuels, which at that time were fuel oil and gas oil (both derived from petroleum). The take-or-pay and the oil-linked price formulas for natural gas have survived as the pillars of gas trade throughout continental Europe until the present, although they are now under attack.

LNG transportation has had a bumpier history than gas pipelines since the technology for gas liquefaction and regasification became available in the 1920s. In 1939, in Cleveland, Ohio, the first industrial installation to liquefy natural gas was built, stocking it in liquid form and regasifying it as necessary to satisfy peak demands. This was called a *peak shaving plant*. That first installation seeded the fears that have dogged LNG for much of its history.

In 1944, the Cleveland plant developed a leak because of a crack in one of the storage tanks. Spreading into the sewers of the city, the gas quickly caught fire, causing fires and explosions and killing more than a hundred people. The government investigation later proved that the accident was caused by the poor quality of the steel in the tank. A low-quality alloy had been used to save top-quality steel for the war effort. It would take years before anyone would try LNG on that scale again. Only near the end of the 1950s was the first gas liquefaction plant for the international transportation of LNG built, in Lake Charles, Louisiana.[10]

In the 1960s, North Africa began to send modest volumes of LNG to Europe from the new liquefaction plant at Arzew, Algeria, and later the plant at Marsa el Brega, Libya. These first limited experiments at internationalizing LNG proved too expensive for Europe, so it would not be until the end of the 1990s that a new LNG export project would take shape (from Trinidad and Tobago to Spain and the United States).

Things went differently on the other side of the world, in the Pacific basin. In the 1970s, riding the waves of fear about oil supplies, Japan moved decisively toward natural gas. Not having its own resources and not being able to reach the principal suppliers by gas pipeline, it had to develop a supply of liquefied gas, quickly and vigorously. The first modest loads of LNG from the United States, then from Brunei, Abu Dhabi, and Indonesia, triggered what would become the most flourishing market for LNG. Today, Japan and South Korea buy more than 50 percent of the liquefied gas marketed in the world.

For all the problems it encountered along the way, the 1970s remain the turning point in the history of natural gas. When that decade opened, natural gas was still at best a downside of finding oil and at worse a real curse. By the early 1980s, natural gas had already started a climb that has not stopped.

Between Boom and Liberalization

Between 1970 and 2008, world consumption of gas tripled from 1 trillion cubic meters to about 3 trillion cubic meters.[11] After having been used mainly for manufacturing and civil uses (e.g., home heating), over the last quarter-century methane burst into the electrical power generation, especially in the last decade. Between 1998 and 2005, more than 50 percent of the incremental demand for gas came from the thermoelectric sector, because of gigantic construction plans for gas-fired power plants.[12] A few easily identified factors lie at the heart of this success.

First, gas benefited from contained prices until the early twenty-first century, which made it competitive with other fuels, leading to the construction of many gas-fired power plants. Second, gas-fired power plants can be equipped with the *combined-cycle technology,* which exploits the heat from the gas combustion to produce more electricity, and *cogeneration,* which uses waste energy to produce heat or steam to be used for energy purposes. This ensures higher efficiency than can be obtained from other fossil fuels (more than 56 percent, compared to 42 percent for the best coal-fired power plants).[13] Gas-fired power plants also have much lower investment costs and fixed costs than coal- or oil-fired plants, and their construction can be adapted more flexibly. Moreover, power generation by gas is the cleanest way to produce electrical power with a fossil fuel, from the point of view of both global pollution and carbon dioxide emissions.[14]

One of the most remarkable features of the boom era of natural gas is the way that the gas market was liberalized in the United

States, Great Britain, and Canada. The United States led the way in liberalizing the gas market, beginning in 1978 by loosening control of wellhead prices (the Natural Gas Policy Act of 1978); allowing consumers to buy gas directly from producers, taking advantage of free access to the transportation system (FERC Resolution No. 436 in 1985); completely deregulating prices, which have since been determined exclusively by market dynamics (the Natural Gas Wellhead Decontrol Act of 1989); and requiring the separation of transportation and sales businesses (FERC Resolution No. 636 of 1992). Great Britain and Canada liberalized their markets in a little more than ten years, beginning in the early 1980s.

What made liberalization possible was essentially one prerequisite: these three countries had ample natural gas reserves and many gas producers, making them practically self-sufficient (historically, the United States and Canada represent a single market). Coupled with large numbers of utilities and industrial consumers, opening the gas markets generated real competition in all segments of the gas chain. At the same time, it allowed the formation of spot markets in which the price was set by the daily meeting of demand and supply, not connected to the prices of alternative fuels such as oil.

A *spot market* is a market for commercial exchanges using short-term contracts (less than two years). These exchanges take place at a *hub,* a meeting point between the different gas infrastructures (production, consumption/delivery, and storage) equipped with a system for transferring the gas from one infrastructure to the other, so that ownership can be associated with the transfer. In the United States, the most important physical exchange point for natural gas, the Henry Hub in Louisiana, sets the price of natural gas for the country. In Great Britain, where a physical hub did not exist, a *virtual hub* emerged, the British National Balancing Point (NBP), encompassing the entire national transportation network. Like the Henry Hub in the United States, the NBP became the price setter for natural gas in Great Britain.

With an abundant supply of domestically produced gas, liberalization allowed the three countries to post lower gas prices than those of continental Europe, where prices continued to be oil-linked on the basis of "take-or-pay" contracts.

However, liberalization had its backlash, too. Prices that had been fairly stable and predictable in the past became more volatile, particularly after the United States and Great Britain began losing self-sufficiency in early 2000s. As a consequence, consumers in those markets became accustomed to the dramatic price swings of a competitive market, sometimes experiencing spikes that made the price of gas (expressed in MBTU) higher than that of oil. Although short-lived, this was a significant reversal of the historical relationship between oil and natural gas prices. It also meant that from 1999 through 2006, average gas prices in the United States were consistently higher than in continental Europe. The same thing happened in Great Britain between 2004 and 2005.

The Rise of Unconventional Gas, the LNG Dilemma, and the Prospect of a "Gas-OPEC"

The high prices of methane in the United States in the early 2000s had positive effects as well. They triggered a silent revolution in the American gas sector—"silent" because apparently no one realized it was happening, until it had happened. As noted by oil and natural gas expert Ed Morse, "The United States used to be considered a country that would eventually suffer a long-term natural gas deficit and be condemned to import supplies; some piped from Canada, other shipped as liquefied natural gas (LNG) from around the world."[15]

This conventional wisdom was shared by the vast majority of experts until 2008, so that most projects of gas liquefaction in Africa and the Middle East were built mainly to supply the United States. Once again in the history of hydrocarbons, prices and technology upset the forecasts.

High methane prices made it economically feasible to apply technologies that had already been used in the petroleum sector to natural gas production, particularly horizontal drilling and hydraulic fracturing. This spurred the development of the large resources of shale gas existing in the United States.

The revolution started at the Barnett Shale field in Texas, which began production in 2000. Since then, there has been a rush to produce shale gas in the United States, leveraging immense deposits such as the Hainesville Shale in Louisiana, and the Marcellus Shale (stretching from West Virginia through Pennsylvania and into New York), which, in the words of Morse, "may contain as much natural gas as the North Field in Qatar, the largest field ever discovered."[16]

As a result, U.S. shale gas production jumped to around 60 bcm (or 2 trillion cubic feet) in 2008, from zero in 2000. In the same year, the whole of U.S. unconventional gas production (including coal bed methane and tight gas) surpassed the production of conventional gas. The dramatic change has been documented in a 2009 report by the Potential Gas Committee (PGC), considered the authority on American gas supplies. In the study, the PGC noted the largest increase in U.S. gas reserves in the forty-four-year history of the Committee: a jump of 35 percent from 2006 to 2008, considering the probable, possible, and speculative resources along with the proven reserves.[17]

New unconventional gas production might actually reverse the declining trend in production in the United States. Shale gas now represents much of the lowest cost gas resource base in the country (excluding proven reserves). One major conundrum concerning future production trends regards prices and costs. If production costs decrease to the 2003 level and if gas prices oscillate in a range of $4–$6, there could be at least 26,000 billion cubic meters (less than 1 trillion cubic feet) of shale gas that could be economically produced. Producible resources may be even more, given the immature state of knowledge of U.S. shale gas. Another problem with shale gas is water. It takes more than 10 million liters of fresh water for a single fracture treatment: A single production well using multiple fractures could need more than 80 million liters. Disposal of the used water is another problem. In the future, concerns about water usage, disposal, and the risk of polluting water tables needed for drinking water could hinder the development of several shale deposits, unless new technological solutions are found.

However, even if shale gas faces temporary setbacks wherever prices drop for too long or if environmental concerns arise, over the long-term it will represent a major consideration in the U.S. quest for cheap, domestic, and low-carbon energy sources. Another important consequence of the U.S. experience with shale gas could be a quest to develop shale gas in other areas of the world. Unfortunately, deposits of shale gas and other unconventional gas are very poorly defined worldwide, and each area outside North America presents specific problems and hurdles that still need careful study.[18]

Finally, U.S. shale gas could play a major role in altering the prospects of global LNG trade, because it is more secure and cheaper than LNG. In other words, most LNG producers will have to carefully review their plans to build up liquefaction capacity, scaling down their forecasts for exporting much of it to the United States. One of the major pillars of LNG development since the mid-1990s has been the need for natural gas in the United States.

When the rush for LNG started in 1996, there were only eight producers of LNG in the world. At the climax of the rush in 2008, there were fifteen, and several scenarios suggested that, by 2020, there could be more than twenty, with a total liquefaction capacity of about 700 bcm, almost three times what we have today.[19]

The 2008 global economic crisis and the upsurge of U.S. gas shale have since clouded these rosy predictions: Many countries have stopped or delayed gas liquefaction projects due to the combination of cost overruns and the decline of gas prices on the international market.

Complicating the already complex scenario of natural gas has been the rise of assertive policies in several producing countries, as they attempt both to protect the value of their exports, and to use them for political and commercial leverage. To many Westerners, the most striking example of such a policy was the running dispute between Russia and Ukraine, which suddenly exploded in January 2006, interrupting the regular natural gas exports from Russia via Ukraine to Western Europe for some weeks. Its background, however, needs to be seen in a historical perspective.

At the time of the Soviet Union, the countries in the Soviet bloc benefited from a political price of gas set by Moscow, paying only a fraction of the price at which Russia sold gas in Europe. After the collapse of the Soviet Union in 1991, in spite of reiterated protests from Moscow and requests for negotiations, the price of gas in former Soviet republics remained between one-fifth and one-eighth of the price paid by the countries of Western Europe. The issue was particularly critical for the relations between Russia and Ukraine. Not only was Ukraine a big consumer of Russian gas, but most of the Russian gas destined for Western European countries, or about 120 bcm per year, ran through the Ukrainian pipeline system. Since the 1990s, Moscow had been accusing Ukraine of illegally withdrawing gas from the pipelines heading to the rest of Europe, and not paying for the regular supplies it provided for Ukraine itself. For many gas experts, thus, the crisis that erupted in 2006 was not a complete surprise, because it was the consequence of a real problem that Russia could not tolerate anymore. Nevertheless, for many observers it was a sign of resource nationalism by Moscow to obtain greater influence both over its former allies in the Soviet bloc and over the many European countries that depended on Russian gas for a relevant part of their energy needs.

This view was apparently supported by the general wave of resource nationalism of the first decade of the twentieth-first century. In the case of natural gas, that wave provoked the first attempts by some large producers to establish a cartel for gas modeled after the Organization of Petroleum Exporting Countries (OPEC). This idea was first proposed by the leadership of the Russian giant Gazprom in the late 1990s, but so far, it has had a hard time getting started. In 2001, a Gas Exporting Countries Forum (GECF) was established, whose participants have only expressed their willingness to strengthen collaboration and coordination, while establishing a working group to find a mechanism for setting gas prices. The most significant developments since then have been the announcement of the creation of a "Big Gas Troika" in October 2008 within the framework of the Forum (Russia, Iran, and Qatar), and the unofficial signing of the Forum's Charter document in December 2008.

The major problem concerning such a cartel will be the operating mechanisms of most gas markets, especially the long-term contracts in which the price of gas is indexed to the price of petroleum. Producers could not manipulate the market value by changing the supply. Other peculiar differences between oil and methane make it even more difficult to establish a cartel.[20]

Furthermore, a gas cartel would risk running into the same problems that OPEC did: lack of discipline among its members and the desire of each to obtain better conditions for itself. Russia and Algeria, for example, have competing interests in Europe, while Qatar would like to take market share in Europe from both countries for its LNG exports.

Nonetheless, a prolonged period of low prices, particularly in Europe, along with a substantial departure from the oil-linked contracts, might create a stronger incentive for GECF to effectively turn into a cartel and establish its own price mechanisms in order to protect its producers' revenues and cover the short-term marginal cost of new production.

Regardless of the potential ability of a cartel to control gas supplies, another phenomenon could affect the availability of gas on the international market: the explosion of domestic demand in producer countries. Many of them are facing unprecedented economic growth and need more electricity in their civil and manufacturing sectors. They also need to reinject gas into petroleum deposits to increase recovery.[21] In 2005, gas reinjection into deposits worldwide totaled about 400 bcm, which corresponds to about 70 percent of the current gas requirement in Europe. About 105 bcm were reinjected in the United States, 80 billion in Algeria, 40 billion in Norway, and 35 billion in Iran.[22]

This exponential growth in domestic demand affects many countries of the Middle East and North Africa, which have considered using nuclear power to sustain their own domestic electricity consumption and free up the available gas for reinjection or export. (Iran is already doing this.) If demand in these countries continues to grow at a sustained pace, the methane available on the international market could prove less than the import needs of the consumer countries.

Thus, several changes are under way in the natural gas market that may significantly alter its patterns in the next decades. Much more than for oil, those will involve delicate geopolitical issues, from the quest for nuclear energy of many Middle East countries to the behaviors of some big producers vis-à-vis their traditional consuming countries. This last problem is of critical relevance for the European Union.

The European Problems with Gas

The European Union depends on imports for most of its gas consumption. Of the 526 bcm it consumed in 2008, around 320 bcm (more than 60 percent) came from a handful of exporters: Russia (39 percent), Norway (28 percent), Algeria (16 percent), and Libya (3 percent). The rest came from other countries, mainly as LNG.

Furthermore, domestic production in the twenty-seven countries of the European Union is dropping at a sustained rate (6 percent per year), which means it could drop by about 90–100 bcm by 2020. How much of this will translate into European demand for imported gas is difficult to estimate. The 2008–2009 economic crisis caused gas consumption in Europe to decrease by about 8 percent.

Most observers think that European demand will grow again after the economic downturn, albeit at a slower pace. Some pessimists think it will never recover, partly because of aggressive targets for 2020 set by the European Union for renewables, energy efficiency, and limits on greenhouse gas emissions.

Nevertheless, even assuming zero growth in demand, by 2020 the European Union will need to import more than 80 percent of its gas requirement. Peak demand for gas (e.g., demand during winter or to back up wind and solar energy) will further complicate the gas consumption puzzle. Redundancy will be needed to respond to erratic consumption patterns throughout the year. This will increase the overall cost of the European energy system.

These problems are aggravated by three peculiar features of the European gas market. First, it is more rigid than that of the United States or Asia, where there is substantial use of nuclear power or coal. Europe cannot count on those two to replace methane in the next decade.

Second, many European countries are not adequately linked to each other, and national interests preclude an easy solution to this problem. The lack of infrastructure means that a country with too much gas cannot resell it to another country. In economic terms, the European gas system is not an integrated market where gas can freely flow from one point to another. The patchwork of different systems and rules increases the rigidity of the market as a whole.

Third, most European gas imports in the future will continue to come from Russia and to a lesser extent from Algeria and Norway.

For years now, the European Union has pursued complete *liberalization* (deregulation) of the gas market, which of itself is a good thing. However, liberalization in Europe has been a controversial issue, with some potentially damaging consequences.

A market is free when all the links in its chain are free, starting with the producer of the commodity that feeds that market. Brussels cannot liberalize the policies of countries like Russia and Algeria, which hold the keys to the present and future European gas market. The only stick that EU authorities could wield is competition among various gas sellers, that is, among those who purchase from the gas oligopolies and sell to European customers. This does not mean that the market has been liberalized. As long as the gas supply for Europe is concentrated in the hands of a few producers who can act as an oligopoly, by definition there cannot be a free market. What good is it to increase the number of sellers in Europe if they all have to turn to the same supplier? Above all, how does this help reduce prices for the end customer or increase the security of the European gas market?

Another target of the European Union and of many European governments has been to diversify sources of supply, especially through LNG or piped gas coming from new producers. Strongly supported by the U.S. administration during the Bush era, this strategy aimed mainly to reduce European dependence on Russian gas.

So far, this support for diversification has led to a flurry of projects to import additional gas via pipelines or LNG. Most of these are still on paper except for the construction of a significant regasification capacity, particularly in Great Britain and Spain. To complicate the picture, traditional exporters such as Russia and Algeria have proposed new pipelines to Europe, leveraging the availability of their gas and trying to keep out potential intruders from the European gas market.

The 2008–2009 economic crisis, the upsurge of U.S. shale gas, and the tightening world market for LNG sowed the seeds of several changes. Together, these three factors have slashed consumption and left Europe oversupplied with gas—including LNG that was initially destined for the United States. In turn, this situation has begun eroding some pillars of the European gas market.

First, the "oil-linked" price formulas in most European gas contracts have become harder to justify, simply because of the ease of switching between natural gas and oil products in heating and power production—as shown by gas expert Jonathan Stern in particular.[23]

The volatility of gas prices during the first decade of this century first obliged consuming countries to pay ever-increasing gas prices that moved according to oil prices. Then the fall of oil prices after 2008 dragged down gas prices too, stressing the finances of many exporters. No one on either side has reason to be happy with the old system, but several producers may have a special advantage in scrapping it.

Oversupply has caused most European buyers with "take-or-pay" contracts not to retire the minimum quantities of gas prescribed by their contractual obligations, but this situation has also curtailed the gas exports of their producers. In turn, this has pushed exporters to start selling gas at discounted spot prices at the different *de facto* hubs of the European market. In doing so, they have taken advantage of the capacity of pipelines left unused because of curtailed demand, entering in direct competition with their customers—the ones bound to them by "take-or-pay" contracts. Thus, Great Britain's NBP has come to be considered a better price setter for natural gas: in 2009, spot prices there averaged half the prices of "oil-linked" formulas.

Direct sales at spot prices would seem to achieve an ambition shared by some (but not all) of the big gas exporters: to eliminate the broker role played by their major clients—the big European "national gas champions"—and supply the European markets directly. Direct sales also signal that they are trying to reduce the advantage of buying LNG, particularly the growing flow coming from Qatar.

Another consequence of the 2008–2009 crisis has been the slowing of infrastructure projects to import more gas in Europe. Although low gas prices and reduced consumption make most of those projects economically unjustifiable, many of them will be pursued with determination—although with uncertain success—for geopolitical reasons.

This entanglement of new factors means that it is still too early to make a serious assessment about the future direction the European gas market will take. Much will depend on the duration of falling gas prices and overproduction, and how the turnaround unfolds. The longer the current crisis persists, the more stress it will place on "take-or-pay" and "oil-linked" formulas, leaving spot prices and fierce competition among exporters to redefine the market. However, if gas prices on continental Europe fall to British levels, they could provoke a rebound of consumption sooner than expected. Such a rebound could impede market penetration by renewables and accelerate the replacement of old nuclear and coal plants with gas-fired ones. In the medium-to-long term, this would make gas use even more widespread within Europe, once again making the security of gas supplies a critical issue.

At the same time, it is not even sure that Great Britain will continue to enjoy lower gas prices than continental Europe. The country may run short of power in a few years, unless it takes drastic and concrete action about energy. Most of its nuclear plants and about half of its coal plants will close in the near future, either because they are too old, or because they are too polluting (in the case of coal plants). Great Britain's hope to develop a substantial renewable capacity in the next eleven years is faltering—in particular, the plan for 33GW of wind energy. As remarked by *The Economist*, "Since coal is too dirty, nuclear plants are too slow to build and renewables are of only limited use, investors are turning towards more gas plants."[24] If this occurs, gas prices in Great Britain will rise again.

In sum, the puzzling European gas scenario holds no certainties. It is likely that we will see unprecedented turmoil and volatility there in the next few years. It is also likely that the European Union will not be able to count on large domestic shale and other unconventional gas resources to stem its declining production, as the United States has done. The geology of Europe presents only small and scattered unconventional gas deposits; production from these probably could not exceed 10 bcm by 2020.[25]

One thing is sure: whatever the scenario, the European Union will depend ever more on imports, and geopolitics will once again play a major role in the European gas market.

Natural Gas as a Transportation Fuel

The technologies to use methane to fuel motor vehicles have existed since the early twentieth century, and the first commercial gas-powered vehicles appeared in the 1940s. Yet, sixty years later, we entered the new millennium with only slightly more than one million natural gas vehicles of all kinds worldwide. The low oil prices of the previous century had discouraged the development of a mass market for these vehicles. Since 2000, their use has grown by more than eight times. Today, about 9.6 million of the 830 million transportation vehicles in the world are gas powered.[26]

Leading the way in this boom is South America. Argentina and Brazil each have more than 1.5 million gas vehicles on their roads, most of them taxicabs. Since 2005, other countries have contributed to the diffusion of such vehicles. By the end of 2008, Pakistan had more than 2 million gas vehicles, Iran 1 million, India 650,000, and China 400,000. According to the International Association for Natural Gas Vehicles, there could be 65 million natural gas vehicles globally by 2020. That would imply using about 400 bcm of gas per year for transportation alone. Today the world consumes about 3,000 bcm of methane every year.[27] Most gas-powered light vehicles use compressed natural gas (CNG),[28] while most heavy-duty vehicles, such as trucks and buses, use LNG. There are also trains that run on natural gas (e.g., one of the four locomotives of the Napa Valley Wine Train).

There are several advantages to using natural gas as a transportation fuel. It is cleaner than most alternative fuels. It is also much safer than other fuels in the event of a spill, because natural gas is lighter than air and disperses quickly when released.

However, there are also several problems connected with its future development. Existing gasoline- and diesel-powered vehicles need to be adapted to use CNG or LNG. First, natural gas cylinders are usually located in the trunk, reducing the space available for other uses. CNG and LNG tanks need to be larger than gasoline or diesel tanks, because of the lower energy density of their contents. The range of a gas-powered vehicle is still less than that of a gasoline or diesel model. Gasoline or diesel engines need to be modified to use methane. Finally, safety concerns have led several cities around the world to limit locations where CNG vehicles can be used and even parked, and certification costs for conversion are quite high in many countries (in the United States, for example). These factors increase the cost and decrease the attractiveness of running a gas vehicle.

Then there are the problems and costs of developing an infrastructure for fuel distribution. Natural gas needs to be insulated and controlled at fueling stations to avoid leaks. On the plus side, gas vehicles can be refueled from existing natural gas pipelines, which makes home refueling possible. Honda and its Canadian partner Fuelmaker have developed a technology sold in the United States under the brand Phill Home Refueling Appliance. Phill can be installed in a garage or outside a home to allow refueling using a residential natural gas supply.[29]

Problems are more significant for LNG, not only because its production costs are higher than CNG but also because LNG requires cooling and cryogenic tanks for storage at fueling stations and in the vehicle to keep it in its liquid state. With the limited diffusion of gas vehicles, these problems have so far discouraged oil and gas companies from developing distribution systems capable of adequately covering a country or even a single area of a country. In turn, this has discouraged consumers from choosing natural gas vehicles, given the difficulty of refueling them. If natural gas prices become competitive with oil prices, especially in certain parts of the world, a new paradigm for natural gas vehicles could emerge, giving credence to the forecasts of the International Association for Natural Gas Vehicles. Considering that natural gas is seen as an environmentally friendly fuel, subsidies and tax credits (already in place in many countries) will probably encourage its future development.

There is another way to use natural gas as a vehicle fuel: transform it into a liquid fuel similar to diesel, which could also be mixed with diesel. The technology for doing this is known as *gas-to-liquids* (GTL). This should not be confused with LNG; while LNG serves only to transport and store natural gas, which is regasified before use, GTL technology permanently converts the natural gas into a liquid. The basic principles and processes for the conversion are the same as those used for coal liquefaction, developed by the German scientists Fischer and Tropsch in the 1920s.

In the 1990s, the future of GTL seemed bright. Several projects were launched or started worldwide. Several automobile and airline companies studied the use of this fuel. Many prototype GTL cars were launched, and in 2007, a commercial airplane, an Airbus 380, first started flying with GTL-based fuel. The bright prospects for GTL seemed to offer some peculiar advantages. In particular, it offered the possibility of exploiting isolated gas fields (too far from markets), and the fuel was compatible with existing diesel engines and distribution systems. However, diffusion of this technology has run into a wall of economic realities.

To repay the enormous initial investment, GTL requires large deposits of natural gas. At the same time, these large deposits must be far enough from end markets or coastal terminals to make local use or export of the gas unprofitable. In other words, GTL is only an option when other upstream and midstream possibilities are not viable. There are very few deposits in the world where GTL would be the best option. Finally, the GTL conversion requires a lot of energy, making it even more expensive. The initial enthusiasm for gas-to-liquids evaporated, replaced by a façade of cautious commitment to develop this technology. The truth is that several of the largest commercial GTL projects have been delayed or canceled.

The Future

For all its problems, natural gas will be a key player in any future energy scenario. It is hard to determine whether methane will rise to second or even first place, overtaking coal or oil. Its future will depend mainly on two factors.

The first stimulus for gas would be the closing of many nuclear and coal fired power plants, which will reach the end of their life cycles over the next twenty years. Many countries affected by this problem are trying to keep the reactors and coal plants in operation. The fact remains that it will be difficult to avoid closing some of the older installations, and at least in the industrialized countries, it will be just as difficult to build new ones. If this scenario plays out, natural gas would be the first choice as a temporary substitute for coal and nuclear power, as the British experience seems to demonstrate. The real issue is that in such a short time-frame, renewables cannot make up for the huge quantity of energy delivered by coal and nuclear power, while fuel oil is not convenient and pollutes more than methane.

The second factor is the timing and effectiveness of public policies aimed at curbing greenhouse gas emissions and local pollution. If those policies are not delayed indefinitely, and if they impose clear and stringent limits on emissions and pollution, natural gas consumption may leap in the next two decades, and natural gas could replace coal as the second largest source of primary energy worldwide by the 2030s.

For sure, gas price trends will feature greater volatility than in the past. Like oil, methane seems destined to become prey to "boom-and-busts" cycles, periods of oversupply and low prices alternating with periods of restricted supply and high prices. Like oil, however, each rise or fall will create the prerequisites for the turnaround. Thus, periods of low prices will push consumption upward while reducing supply, creating the conditions (high prices and tight supply) for investing in the development of new and more expensive production—such as unconventional gas. The trend will not be linear, but in my view, the long term direction of natural gas points to a final landing. However the future unfolds, I think that at some time in this century, natural gas will displace its powerful and respected cousin, oil, as the number-one source of energy for humanity.

Statistics on Natural Gas

Table 3.1. Proven natural gas reserves and reserve life (Top 10 countries and world total, 2008)

Country	*Proven reserves (trillions of cubic meters)*	*Reserve life (Years)*
Russia	44.9	72
Iran	29.3	252
Qatar	25.2	328
Saudi Arabia	7.5	95
United States	6.7	12
United Arab Emirates	6.4	158
Nigeria	5.3	163
Venezuela	4.8	186
Algeria	4.5	52
Indonesia	3.2	46
World total	**180.9**	**59**

Table 3.2. Natural gas production (Top 10 countries and world total, 2008)

Country	*Production (billions of cubic meters)*
Russia	621
United States	582
Canada	167
Iran	116
Norway	99
Algeria	86
Saudi Arabia	78
Qatar	77
China	76
Netherlands	76
World total	**3,058**

Table 3.3. Natural gas consumption (Top 10 countries and world total, 2008)

Country	*Consumption (billions of cubic meters)*
United States	669
Russia	467
Iran	117
United Kingdom	96
Japan	96
Germany	88
Italy	86
China	81
Canada	80
Saudi Arabia	78
World total	**3,010**

Table 3.4. Principal uses of natural gas, by sector worldwide (2008)

Sector	*Share (%)*
Electricity	37.6
Heat	3.6
Other energy transformation sectors	10.3
Manufacturing	17.3
Civil and agricultural	23.8
Transportation	0.4
Nonenergy uses	5.7
Consumption and distribution losses	1.3

CHAPTER 4

The Quest to Eliminate Carbon Dioxide

Because fossil fuels will continue to play an overwhelming role in satisfying the energy needs of the world for a long time, we must deal with the damage they cause to the planet and its atmosphere. We can lower their local and regional polluting potential (sulfur, particulates, etc.), and in recent decades much progress has been made. But it will take a large collective effort in scientific and technical research, supported by laws and long-term restrictions, to continue reducing the release of dangerous substances.

Limiting carbon dioxide emissions is more complex. About 60 percent of the CO_2 that humans produce comes from burning fossil fuels.[1] Since the end of the twentieth century, the greatest hope for eliminating the CO_2 has been a process that involves its capture and geological storage: *carbon capture and storage* (CCS), also known as *carbon capture and sequestration*. While this option is worth pursuing whenever possible, it also presents several limitations. CCS will thus represent an important weapon in the fight against climate change, but we will need to look far beyond it to deal with the overall problem of CO_2.

The Geological Storage of Carbon Dioxide

Some CCS technologies are already available, thanks mainly to tests conducted by the petroleum industry since the 1970s. These technologies separate the carbon dioxide before or after burning coal, petroleum, or natural gas. The CO_2 is concentrated and compressed, then transported by gas pipeline to an underground deposit. This is usually a deposit from which it is no longer possible to recover crude oil or natural gas (technically exhausted) or a *salt aquifer* (a porous geological formation saturated with salt water). The CO_2 is injected into its new coffin and monitored to keep it from escaping.

In general (but not always), geological deposits where carbon dioxide can be stored must be at least 2,000 feet (800 meters) deep, consist of porous rocks or minerals to absorb the CO_2, and be surrounded by strata of impermeable rock, which prevent the CO_2 from escaping. The great depth is necessary to keep the CO_2 away from water tables that could be used for drinking water and to store it at a pressure that will keep it close to its liquid density, rather than as a gas. In this *supercritical state,* the possibility of the stored CO_2 escaping is very remote.

The first commercial-scale CCS project dates back to 1996. Located at Sleipner in the North Sea about 150 miles (250 kilometers) from the coast of Norway, it is still in operation, managed by the Norwegian oil company Statoil with other petroleum companies. Sleipner sequesters more than one million tonnes of CO_2 each year. There are other operational commercial projects in Canada (Weyburn) and Algeria (In Salah), and the principal players are always petroleum companies.

There are other ways to sequester this greenhouse gas. Since the 1970s, mainly in the United States, carbon dioxide has been reinjected into petroleum deposits to increase the production of crude oil. This advanced technology makes it possible to extract more petroleum than would be possible with conventional methods. Currently on average, we can extract no more than 35 percent of the crude oil contained in known deposits worldwide, and achieving even this figure requires injecting water or natural gas into those deposits. We can overcome this limitation with carbon dioxide injection and other technologies.

Carbon dioxide can also have the opposite effect on oil or gas wells. Depending on the geological structure of a deposit, its internal fractures, and other factors, injecting carbon dioxide could irreversibly damage the ability of the deposit to produce oil or gas. In other words, using CO_2 to increase the productivity of an oil deposit is not a general option, but must be assessed case by case. CO_2 can also be used as a cushion gas in methane storage deposits. Here, too, the feasibility depends on the characteristics of the underground deposit.

The United States already has 1,500 miles (2,500 kilometers) of pipelines devoted to moving CO_2, and each year 30 million tonnes of carbon dioxide are injected into oil and gas deposits. That is approximately the annual CO_2 output of ten million American cars or a single old 1-gigawatt (GW) coal-fired power plant.

Sequestrating carbon dioxide underground would be a great boon not only for oil and gas extraction but also for industries that release large quantities of CO_2, such as electric utilities. Making CO_2 disappear from the production cycle would allow us to rethink the use of coal for power generation, thus increasing the flexibility and the security of national energy systems.

CCS has an important future, but the geological storage of carbon dioxide must still overcome some problems before becoming a real option on a broad scale.

The Problems

CCS presents three challenges to those who would see it used extensively: its cost, the fear of CO_2 escaping, and the fear of a geological disaster at the storage site. Capturing and sequestrating a ton of CO_2 today is very expensive. The cost varies widely, depending on the conditions under which this greenhouse gas is captured. The capture phase accounts for 70–85 percent of the total cost of CCS, compared to a modest 5–10 percent for storage and 10–20 percent for transportation. The reason is simple. The sources that emit carbon dioxide are widely scattered, and each of them generally produces small amounts of this greenhouse gas. For example, it takes two million automobiles to produce the same volume of CO_2 as a 1-GW coal power plant with the latest scrubbing technology (about 6 million tonnes of CO_2 per year). Furthermore, most sources generate CO_2 with other gases and substances, which must be separated out to obtain enough carbon dioxide to justify a CCS project.

This has two implications. First, it is unlikely to be cost-effective to capture the CO_2 from many, small, widely dispersed sources. Second, carbon dioxide from a large point source (e.g., an oil or gas field, a refinery, or a coal-fired power station) must be treated before it can be economically captured. Capturing CO_2 rests on an apparent environmental paradox. As the production and concentration of CO_2 rises, the cost of capturing it goes down. This means that capturing one ton of carbon dioxide produced from burning coal costs less than capturing the CO_2 from natural gas. However, the picture reverses if we correctly consider the cost of CCS for each MW of electricity produced. According to the analysis by MIT professors Hamilton, Herzog, and Parsons, CCS would add $40/MWh hour to the price of coal-fired generation and cost $60-$65/ton of CO_2 avoided, while it would add $30/MWh to the price of gas-fired generation and cost $85/ton of CO_2 avoided.[2]

The costs per ton of CO_2 avoided are much higher than those currently registered on the European market for emission rights, which as a consequence of the economic downturn fell in 2009 to a range of $10–$15/ton (from nearly $30 in July 2008). They are also much higher than those appearing in the long-term budget submissions prepared by the U.S. administration, which anticipate a price of $20/tonne for carbon emission rights.

The high cost of CCS can undo well-meaning national policies to deal with greenhouse gases. If governments were to impose CO_2 penalties, the cost of CCS could make it more cost-effective for large CO_2 producers to pay penalties than to cut emissions. If a cap-and-trade system to limit CO_2 emissions were in place, with market forces determining the price of carbon allowances, the cost of CCS could force CO_2–intensive industries (steel, concrete, oil refining, etc.) to relocate to countries where the laws are looser or nonexistent.

Work is being done to reduce the costs of capturing carbon dioxide from coal combustion, mainly by gasifying coal before burning it (the same thing can be done with oil). Gasification is a technology that is already widely used to generate electricity. It produces not only a relatively pure and very concentrated flow of CO_2 but also hydrogen. However, there are still many complications and cost obstacles to deploying this technology commercially.

There are still no zero-emission, coal-fired power plants with CCS. There has been no shortage of planned projects, heralded by press releases and aggressively touted by their proponents, but most of them fell apart after the initial hype. Others are still in limbo.

As the prospect of testing a large-scale, zero-emission, coal-fired plant comes closer, other issues complicate the issue of carbon capture and storage. Some people are worried about the risk of the carbon dioxide escaping from its underground coffin and returning to the atmosphere or, worse yet, getting into underground water tables. Experiments conducted so far have shown that this is a very unlikely hazard. The storage projects already in operation have reported escapes that are so close to zero as to be irrelevant. Scientists and the petroleum industry have also developed sophisticated models to forecast the interaction between the injected CO_2 and the reservoir destined to contain it, enabling continuous monitoring of the geological coffin.

The choice of the right storage deposit in the right area (most important, far from seismic areas or volcanoes) and careful monitoring are key to avoid what happened in Lake Nyos, Cameroon, in 1986. Lake Nyos sits in an active volcano crater. The CO_2 produced by the volcano seeps into the lake above it, where the pressure of the water traps the CO_2 on the lake bottom. On August 21, 1986, the lake suddenly released a large cloud (some estimates say about 300,000 tonnes) of carbon dioxide (an *overturn*). Being heavier than air, the CO_2 slid down the volcano into the surrounding valley, asphyxiating about 1,700 people and 3,500 livestock before dissipating. The cause of the overturn in the lake was probably a mild earthquake or a volcanic tremor causing an underwater slide. This incident is often touted as evidence of the dangers of storing carbon dioxide. However, it was a natural disaster caused by a unique set of conditions that exist only in three volcanic lakes in Africa. A carefully designed CCS project would not come close to replicating those conditions.

The storage capacity of the subsoil of the Earth is immense, and it offers the possibility of storing several trillion tons of CO_2, particularly in salt aquifers. To give an idea, the salt aquifer of Sleipner alone has a capacity of 600 billion tonnes (the CO_2 released each year by human activity is a little less than 50 billion tonnes).

Sequestrating at least part of the carbon dioxide produced by human activity remains one of the important objectives of climate policy that each country must pursue. The Intergovernmental Panel on Climate Change (IPCC), established by the United Nations and the World Meteorological Organization, has done so much to raise awareness about climate change that it was awarded the Nobel Peace Prize in 2007. The IPCC has great expectations for CCS. Unfortunately, most countries have no standards or regulations on the subject, so realizing projects of this type remains either virtually impossible or extremely complex.

Looking beyond CCS

Even if more countries were to embrace CCS in their policies, it would be an illusion to think of capturing and storing all the CO_2 produced by human activity in the world. Only a small part of that CO_2 can be effectively imprisoned in the subsoil.

There is a big problem with CCS, which is often overlooked. I call it the *relative scarcity of carbon dioxide.* Such a statement deserves an explanation.

Given that the costs of capturing CO_2 are very high, it follows that we need to minimize the costs of transporting and storing it. In turn, this requires geographical matching of each large carbon dioxide source with a place where the CO_2 can be stored. Otherwise, a long pipeline would be required, making the project unaffordable.

The problem is that most large sources of CO_2 are very far from places where the gas can be geologically sequestrated. In other words, CO_2 is mainly produced where it should not be, and suitable geological deposits are where CO_2 is scarce (at least not in large, concentrated streams that would be affordable to capture.

The geographic mismatch between the sources of large streams of carbon dioxide and the geological deposits where it can be stored has not been well studied, either globally or by country. Yet there is much to suggest that the problem is relevant, because it increases the cost of an already expensive activity. For this reason, and for the problems just outlined, my expectations about the future role of CCS are more guarded than those of the International Energy Agency (IEA), which estimates that CCS should be responsible for almost 20 percent of the total greenhouse gas emission reductions by 2050. The IEA figure includes the other greenhouse gases such as methane.[3]

CCS is only a partial solution to the general problem of eliminating CO_2. We must pursue CCS wherever possible, but it would be a mistake to focus on it exclusively. There are other technological frontiers that deserve to be explored, even if they may require time to deliver significant results.

One line that deserves to be pursued is extracting the CO_2 from raw materials processing to obtain carbonate rocks using chemical substances and reactions. The rocks could then be used for construction and other uses. Another line of promising research is to use specific solid materials that may adsorb or absorb CO_2, like a sponge, and store it permanently.

The most advanced laboratories in the world are studying systems that border on science fiction trying to solve this problem, trying to find ways to accelerate processes that nature took millions of years to perfect, or exploring new materials and complex chemical and physical processes to entrap carbon dioxide.

Whatever the difficulty, acting directly on CO_2 emissions from fossil sources seems indispensable to me, for two reasons. The first is that, considering our present and future prospects for alternatives to fossil fuels, the latter will continue to dominate the energy scene of the world for the coming decades. Therefore, reducing the direct release of carbon dioxide will be fundamental.

The second concerns the energy policies of such countries as China and India. During the period covered by the Kyoto Protocol, these two countries will have built almost eight hundred coal-fired electrical power plants, which will produce between three and five times as much carbon dioxide as the Protocol set for the worldwide reduction of CO_2. Because no one is in a position to stop the gigantic energy momentum of these two countries, the only feasible objective is to temper it with technological solutions that can restrain the carbon dioxide that their drive will produce.

While China seems eager to develop its own solutions or to buy technologies worldwide for dealing with the problem, India and other emerging countries are worried about the financial burden that eliminating carbon dioxide may put on their economic development. It follows that the post-Kyoto era should find a mechanism that supports technology transfer to those countries, and financial assistance to ease the costs they will face. This is part of the bill that the industrialized countries may need to pay for having guzzled energy and cranked out greenhouse emissions so far, if they want to be credible when asking newcomers to go on a diet for the sake of our planetary health.

PART TWO

The Alternatives

CHAPTER 5

Nuclear Power: Renaissance or Decline?

Arguably, nuclear power stands out as the most credible modern complement to fossil fuels. It has managed to establish itself relatively quickly (within about fifty years). In that time, its share among primary energy sources has grown to more than 6 percent overall, and it covers more than 16 percent of the electricity consumption of the world. It could potentially provide very large amounts of energy without carbon dioxide emissions. Electricity costs from nuclear power seem close to those from fossil fuels. Nevertheless, some serious unknowns cast doubt on the ability of nuclear power to provide a convincing response to our energy problems in the near future. The initial investment costs are very high. The return on investment can be threatened by delays in obtaining permits, accidents, cost overruns in plants construction, and low electricity prices.

Plant safety remains a serious concern in the public mind, even if there has been no discernible impact on public health from Western reactors, and the very high capacity factor at which Western nuclear reactors work (around 90 percent) suggests excellent operational patterns. Permanent storage of highly radioactive waste is still an unsolved problem after decades of studies, experiments, and simplistic hypotheses. More than anything else is the problem of simple fear.

However irrational or exaggerated, the multifaceted fear of nuclear power (fears of reactor accidents, of radioactive waste, of environmental contamination, of terrorist attacks, etc.) is a formidable obstacle to its development. It is an economic paradox that only the poorest countries or developing countries can build a nuclear reactor relatively easily. Except for Japan, it is almost impossible in most industrialized countries.

Today, a consensus seems to be growing in favor of nuclear power, although there are sharp divisions on details. In most European countries, for example, if a referendum were held in which the voters were to indicate simply yes or no to nuclear power, the yes vote might prevail in many countries. However, a referendum asking voters to have a nuclear power plant *in their own town* would almost certainly be turned down by a wide margin. This does not seem to be the case in the United States for communities that already have nuclear plants. Most seem fine with adding another and indeed, that is where the next reactors would go.

NIMBY ("not in my backyard") is the real problem for nuclear power, so much so that it has become impossible to build new plants or to keep existing ones in operation. Since the expansion in the 1960s and '70s, nuclear power development has stagnated and declined. Notwithstanding a potential rebirth proclaimed recently by important international publications, most notably *The Economist*, the contribution of nuclear power to the production of electricity in the world is increasing in absolute terms, but it continues to decrease as a percentage of total electricity produced.[1]

Yet it is difficult to imagine solving the energy equation without some contribution from nuclear power.

Rise and Fall

After the first commercial reactors came on line during the 1950s in Great Britain, the United States, and the Soviet Union, nuclear power grew at a steady pace for almost thirty years.[2] By 1980, there were 243 nuclear power stations in the world, with a total capacity of about 140,000 megawatts (MW). Most of them (162) were built during the 1970s, thanks in part to the first oil shock, which not only made nuclear power profitable but also showed it to be more secure than oil from the point of view of supply.

For these reasons, not even the first big nuclear accident (Unit Two at Three Mile Island in Pennsylvania in 1979) halted the race, though it did begin to undermine faith in this energy source.[3] Irrational fears about the uncontrollable effects of manipulating the atom to obtain energy were already gaining traction, feeding popular fiction like the film *The China Syndrome* (1977), based on the idea that the core of an atomic power plant could implode and perforate the Earth to emerge on the other side of the globe. With the second oil shock (1979–1980), the nuclear power boom continued into the early 1980s, featuring the construction of another 176 reactors. By 1990, there were 419 nuclear power plants in the world, with a total capacity of 325,000 MW.

Then in March 1986 came the disaster at the nuclear plant in Chernobyl, Ukraine. The accident caused direct deaths and the winds carried a radioactive cloud over Europe. It left an aftermath of sick and deformed victims and environmental devastation. Today we know that human error by the operators was central to the accident but also that the tragedy at Chernobyl was caused by the critical state of the plant, built with obsolete technology and lacking safety features that would have prevented a similar accident in the West. Nevertheless, the world was so shocked that this one accident almost completely stopped the clock on nuclear power.

Petroleum was also involved. In the same year as the Chernobyl accident, high crude oil prices, which had supported the growth of nuclear power for years, collapsed in an excess of supply that could no longer be absorbed. Suddenly, nuclear power seemed to be not only a terrible scourge but also no longer a cost-effective alternative to oil and gas at bargain prices.

The development of nuclear power slowed dramatically. Today in the world, there are 438 nuclear power plants (only another 19 since 1990), distributed in thirty countries with an installed capacity of about 370,000 MW.[4] The United States has the most power plants, more than a hundred, but it is in France that nuclear plants provide the largest share of electrical power (more than 78 percent). Currently thirty plants are under construction with a total potential of 24,000 MW. There are plans for 120 new plants in various countries.

In the United States, where new plants have not been built for twenty-five years (although several have been completed recently after a very long time since their construction began) there is a strong push to extend the operating life of the reactors. This can be done by replacing some components (e.g., steam generators and pressure piping), modernizing some instrumentation, and thoroughly checking the condition of the plant. More than 50 of the 104 active reactors in the country have received twenty-year operational extensions from the U.S. Nuclear Regulatory Commission.

A new power plant is under construction in Finland, the first in Europe in many years. Meanwhile, another construction site is opening in France. There are discussions in Eastern Europe about renovating or expanding various installations, and many reactors are under construction in emerging countries, especially India and China.[5] Nuclear power represents an ever-growing aspiration in many countries of North Africa and the Middle East, including those holding large reserves of oil and gas.

Alongside this reawakening of interest, there are strong antinuclear currents opposed to any further development. Their arguments have powerful appeal: the intrinsic insecurity of such a potentially devastating energy source; the threat of terrorist attacks or blackmail; the problem of handling radioactive waste; the underestimating of real costs (mainly the failure to consider the cost of closing plants and waste disposal, but also repeatedly underestimating the initial construction costs of new plants); and the risk of misuse of fuel enrichment and reprocessing facilities by countries known for their aggressive intentions.

Navigating between the pros and cons of nuclear power is not easy. The subject itself is technically complicated, and both sides tend to ideologize much of the debate. To find an objective point of view of the prospects for this energy source, it would be good to take a small journey to see what nuclear energy really is, how it works, and what its advantages and disadvantages are on a technical level.

Nuclear Power: What Is It and How Does It Work?

All the nuclear energy that the world uses today and could use in the coming decades comes from the *fission* of the nucleus of particular atoms, that is, from bombarding it with neutrons causing the atomic nucleus to split into two (or, rarely, three) parts plus neutrons. This splitting liberates a tremendous amount of energy.

In almost all cases, the element used for nuclear fission is uranium,[6] which is found in nature as a mixture of different isotopes, mainly uranium-238 (U-238) and uranium-235 (U-235).[7] In the natural mixture, U-235 comprises only 0.7 percent of the total. This is a problem, because in today's thermal reactors (except in Russia) only the U-235 is *fissionable* (meaning that it can be bombarded and used for a nuclear reaction).[8] For this reason, in most reactors existing today the uranium must be *enriched,* increasing the percentage of U-235 at the expense of the U-238.[9]

Before enriching uranium, however, a long process of transformation is required. Uranium ore is first mined as a raw material, and then is leached with sulfuric acid to remove uranium and get uranium oxide, also known as *yellow cake*. The latter is transformed into a gas to start the process of enrichment, at the end of which the basic material for a nuclear reaction is ready.

To operate a nuclear reactor for civilian purposes (i.e., to produce electricity), the enrichment should achieve a percentage of U-235 of 3 to 4 percent.[10] An atomic bomb, by comparison, would generally require enrichment to greater than 90 percent U-235. Alternatively, the same weapon could use plutonium-239.[11] Technically, a nuclear weapon could be built with substantially less enrichment, but at the expense of having to increase *critical mass* (the amount of material needed to sustain the reaction); heavy bombs are hard to handle, conceal and deliver. *Depleted uranium* is what is left over from the enrichment process. This uranium is nearly pure U-238, which is *inert* (not radiologically dangerous) and very heavy, being 1.7 times denser than lead. Among other things, it is used for weapon projectiles requiring a high penetrating power.

Certainly, the principal advantage of uranium is its enormous energy density. One kilogram (2.2 pounds) of enriched uranium has the same energy content as 1,800 tonnes of crude oil, 2,600 tonnes of coal, or more than 2.1 million cubic meters of methane. This means that the atom meets the fundamental prerequisite for satisfying the great energy needs of today's society and inevitably of society in the future.

The very high energy density of uranium translates into very high power density, meaning that the amount of the mineral needed to produce electricity is very modest. With some approximating, a 1-GW nuclear power plant needs about 25 tonnes of enriched uranium each year, which can be held in a space as small as a railway car or a freight container.

Such small numbers can be misleading, though. Those 25 tonnes of enriched uranium require the transformation of more than 200 tonnes of uranium oxide (yellow cake), which in turn are produced by mining 25,000–100,000 tonnes of uranium ore, depending on the uranium concentration.[12]

Today, proven reserves of uranium total about 2 million tonnes, enough to feed existing reactors for another fifty years at the current annual production rate of about 40,000 tonnes. However, as with oil, gas, and other natural resources, the definition of proven reserves can be deceiving. The proven reserves include only uranium that is economically extractable for less than $40 per kilogram.

Uranium oxide prices at less than $130 per kilogram would add another 11 million tonnes to the proven reserves—a figure which could increase by an order of magnitude at twice that price. Of course, fluctuations are always possible; the spot price of uranium has gone from an average of $13 per kilogram in 2001 to about $100 in 2006 and well above $200 by May 2007. However, such fluctuations were essentially caused by problems in big uranium mines in Canada and Australia and the peculiarities of the nuclear fuel market. Eventually the price dropped back and averaged about $80 per kilogram during most of 2009. Like other minerals, the future availability of uranium depends on demand, technology, and price. Thus, it is reasonable to expect that fissionable fuel will not have a scarcity problem in this century.

Uranium is not distributed evenly throughout the planet. Even though small deposits can be found almost everywhere, only three countries—Australia, Canada, and Kazakhstan—hold more than 60 percent of the known reserves and are its principal producers.

Besides uranium, we could theoretically use other nuclear fuels, such as thorium, which is much more abundant than uranium. However, thorium based reactors are unlikely to be commercialized any time soon, especially with cheap uranium available, mainly because of the higher cost of fuel fabrication compared to the standard uranium, technical problems still to be solved in the fuel cycle, and some concern over the possible use of spent fuel for weapons.[13]

Along with uranium and thorium, plutonium could also play an important role in producing nuclear power, besides its use in atomic weapons. Plutonium is obtained artificially by using uranium in *closed-cycle reactors,* in which the irradiated fuel is reused. Suitably treated and purified, plutonium from this process becomes a nuclear fuel, but it can also be diverted to make nuclear weapons.[14]

By contrast, in *open-cycle plants,* the waste (irradiated fuel) is disposed of and then treated for storage. The choice of open- versus closed-cycle technology becomes a balancing act of conflicting issues: weapons proliferation, waste disposal, and fuel efficiency.

Nuclear *fusion* remains the Holy Grail of atomic energy. Whereas in fission heavy elements are broken apart, in nuclear fusion, the nuclei of light elements (such as hydrogen) are combined and fused. The mass of the fused atoms is slightly lower than the sum of the individual atoms before fusion. The difference is called the *mass defect*. The mass defect is released during nuclear fusion as an immense quantity of energy: consider the orders of magnitude in Einstein's famous equation $E=mc^2$.[15] Nuclear fusion is occurring constantly in the universe. For example, fusion generates the energy of the sun and the stars. We still cannot produce fusion under controlled conditions, however, except in a laboratory for infinitesimal fractions of a second.

Compared to nuclear fission, fusion has a great advantage: It does not produce long-term radioactive wastes. A nuclear fusion power plant would produce mainly helium-4 (He-4), an isotope of helium that is an inert gas and not radiologically dangerous. For this reason, fusion is the focus of major international research projects and is currently our only promise of truly clean nuclear power.[16]

Unfortunately, we are still very far from having nuclear fusion technology. In the coming decades, we can count only on nuclear fission with its various technologies, so I will concentrate on that in the rest of this chapter.

The energy produced during nuclear fission in existing reactors heats a liquid (water), which flashes into steam. The steam powers a turbine connected to a generator. In most cases, the heat exchange takes place using two separated fluids, which also serve to cool the reactor.

Most reactors in the world are either *pressurized water reactors* (PWR) or *boiling water reactors* (BWR).[17] These two big families of reactors are based on two different ways of cooling the reactor and generating steam. In the PWR, the refrigerant (water kept under pressure so it will not boil) coming hot from the reactor heats a secondary fluid, which vaporizes and goes to the turbine. In the BWR, the heat of the reactor turns the refrigerant into steam, which is sent directly to the turbine.

Other lines of nuclear reactors are classified by the type of *moderator*. This is the material (usually graphite, light water, or heavy water) that slows the neutrons coming out of the fission reaction with high kinetic energy (i.e., at high speed) and stabilizes the temperature of the reactor core at a few hundred degrees. This makes the neutrons more effective at causing additional fissions.

A nuclear reaction is called a *chain reaction* when there are as many neutrons interacting with the fissionable nuclei as there are fissions from which they originated. When this condition is satisfied, the nuclear reaction supports itself. The most important factor for this to occur is the speed of the free neutrons that populate the reactor core. This is where the moderator comes into play, slowing the neutrons issuing from the fission before they interact with other fissionable nuclei. This stabilizes the temperature of the reaction. For this reason, the moderator has an essential role in the nuclear reaction. Not by chance, the legendary Nazi research into heavy water was aimed at finding the necessary moderator to construct a plutonium atomic bomb, in order to avoid the need to enrich uranium.[18]

The best-known classification of reactors is based on the generation to which they belong. Those designed and built before the 1970s belong to the first generation. Second-generation reactors include mainly light water reactors, which were built beginning in the 1970s and are still in operation.

The third generation comes from optimizing the current light water reactors. These include advanced reactors such as the European Pressurized Water Reactor (EPR) under construction in Finland,[19] the AP1000 (Advanced Passive) reactor, and the Advanced Boiling Water Reactor (ABWR) being developed in Japan. The third-generation reactors will be operational in a few years.

The fourth generation will probably not be available until well after 2040. Theoretically, those reactors will provide very competitive prices, greatly increased safety, minimal radioactive waste (especially waste with a long half-life), and greater protection against the risk of nuclear proliferation and attacks. However, hoping to achieve all these targets together calls for a large dose of optimism.

Certainly one of the great advantages of nuclear power is that it does not emit carbon dioxide in producing electricity (if we do not consider the mining of the uranium ore), or sulfur and nitrogen oxides, which contribute to acid rain. From this point of view, nuclear power is certainly friendly to the environment and climate. Unfortunately, on the other side of nuclear power are its notorious radioactive wastes.

Wastes are not all the same. The vast majority of them consist of low-level radioactive waste, whose radioactivity decays after a few hundred years. They can be stored in surface or underground deposits (a hundred feet or so below ground), where they can be kept under control for at least three hundred years.

Highly radioactive waste represents only one-twentieth as much by volume, but its radioactivity persists for thousands of years.[20] For this reason, after a few decades of cooling in surface deposits, they must be inserted into a deep underground deposit in special formations (clay, salt, granite) that will trap them for geological periods. Today there are no sites of this type for civilian use in the world. There is one American site for military use, called the Waste Isolation Pilot Plant (WIPP), located in New Mexico and operational since 1999, but it does not handle high-level radioactivity waste.

Some waste can be treated to separate out new fissionable material for reuse in a reactor, but also for use in atomic weapons. Such are the issues that surround plutonium from closed-cycle reactors, and *breeder reactors,* which make more fissionable fuel than they consume by *fertilizing* the nonfissionable part of their fuel elements.

This brief trip through the workings of a nuclear power plant must end with another of the great problems lacking a sure solution: the "death" of a reactor. Once its planned life cycle ends, a reactor must be stopped and then dismantled so that the site can be restored to its preexisting condition. In technical jargon, this is called *decommissioning*. The delicacy and the complexity of the operation make this phase extremely expensive and complicated.

Since the use of nuclear power to make electricity began, about 110 reactors have been shut down. However, according to the International Atomic Energy Agency (IAEA), by the end of 2005, only eight power plants had been completely dismantled and their sites reclaimed for other uses.[21] A few others have been completely dismantled in the United States and elsewhere, but the vast majority of those shut down still face a very long dismantling process.[22] The unforeseen problems and costs have proven to be much higher than originally estimated.

This is the dark shadow over our near-term future, when most of the nuclear power plants in operation today will reach the end of their useful lives and need to be shut down (on average in forty to sixty years).[23]

The Unsolved Problems

Over the last fifty years, nuclear power research has benefited from public financing in the industrialized countries, leaving only 10 percent of the total funding to other alternatives to fossil fuels.

This immense effort has noticeably improved many aspects of nuclear engineering and technology, but it has not dissipated the principal clouds overshadowing the field. Let us consider them in an orderly fashion.

Although the fixed costs of building a power plant are high, its variable costs (the cost of daily operations, including uranium fuel) are rather low. New-construction reactors can last sixty years, twice the life span of previous models. Distributing the revenues and expenses over a much longer period would make nuclear power more competitive in generating electricity.

This will not make the cost of an "atomic" kilowatt-hour equivalent to a kilowatt-hour from coal or gas. According to MIT, the overnight construction cost for a nuclear power plant in 2007 dollars is $4,000 per installed kilowatt, which would put the cost of a 1000-MW power plant at $4 billion.[24] The same power plant fired by coal would cost $2.3 billion while a gas-fired plant would cost $850 million.[25] By definition, the engineering assessments behind these estimates did not consider delays obtaining permits and other unforeseen expenses. Such delays are the order of the day when building a power plant of any type. For example, in the MIT calculations, the expected time to build a nuclear power plant was only five years; in many cases, changes in regulations, litigation of various sorts, and market downturns often stretch that to fifteen years.

Coal- and gas-fired power plants also have construction cost overruns, but these are much smaller than the overruns at nuclear power plants. MIT did point out that data about power plants built recently have not been transparent or reliable.[26] In any case, what counts in these assessments is not the particular price tag of each project, but the ratio of the costs of the different types of power plants. We can use that ratio even if the underlying costs are underestimated. Using this logic, Cambridge Energy Research Associates calculated the costs of building power plants in the United States during the first quarter of 2009. All costs had increased significantly, reflecting inflation in the cost of concrete, steel, and engineering services. Consequently, the cost of a 1,000-MW nuclear power plant had escalated to $4.7–$5.5 billion, while the same plant fired by coal or natural gas grew to $2.7–$3.3 billion and $0.8–$1.0 billion, respectively.[27]

Factoring in operational costs, the heavy initial price tag of nuclear power comes down noticeably. The total cost per kilowatt-hour produced by a nuclear power plant operating for forty years would be 6.7 cents, compared to 4.2 cents for a coal-fired plant and 4.1 cents for a gas-fired plant.[28] MIT estimated that not even a carbon tax of $50 per ton on carbon dioxide emissions from coal- or gas-fired plants would make nuclear power more economical.[29] On the other hand, extending the forty-year life span used by MIT to sixty years improves the price of nuclear power by another few cents.

At this point, such estimates are theoretical. Historically, the real costs of developing nuclear power plants have always proven to be much higher than the planned costs. The meager data we have on nuclear power plants under construction today indicate that the budgets on which the work was started were greatly underestimated. For example, the initial budget of the Finnish power plant has doubled in a few years. This raises again the question whether nuclear power is competitive in terms of actual costs.

These uncertainties are critical for those who must decide whether to invest in nuclear power in a competitive market, that is, without the special protections and guarantees provided by governments. The first civil nuclear era in the United States was heavily supported by the Price-Anderson Act (1954), which greatly reduced private liability by guaranteeing public compensation in the event of a catastrophic accident in commercial nuclear-powered electricity generation.[30] After World War II, the U.S. nuclear industry was also heavily financed by public funds to jump-start its massive development.

In today's competitive market, faced with very high initial costs, uncertainties about the permit process, the risk of future accidents, or reactions by local communities, it would be nonsense for a private investor to take a leap in the dark with a nuclear power plant. On the other hand, we should not forget that one of the most formidable obstacles to developing nuclear power in most of the world has been the time that passes between the decision and the realization of the power plant.

The planning, permit, and construction phases of a nuclear power plant have always required much longer times than other electrical power plants.[31] In countries that possess the entire infrastructure necessary to realize a nuclear initiative, the period between the political decision to build and the delivery of electricity to the electrical network has varied between seven and fifteen years, but these are historical data, which have little meaning.

Between 2001 and 2005, sixteen reactors were completed in an average of 109 months.[32] Add to this the time needed to obtain a permit for the sites, prepare the environmental impact statements, obtain the consent of the local communities, and complete the connection to the electrical network, the actual time on average becomes about 200 months, except in Japan and China.[33] Clearly, timelines of this sort are incompatible with the idea that nuclear power could play a significant role in meeting our energy and environmental needs in the near future.

The cost and safety aspects of treating waste and decommissioning shut-down power plants make things even worse. In the twenty-seven-member European Union, nuclear wastes are growing by 40,000 cubic meters per year, or about 100,000 tonnes.[34] The accumulation of highly radioactive wastes is already about 17,500 tonnes, with an annual increase of more than 1,700 tonnes. Together, these quantities would make a building with a footprint of 9,000 square feet, 100 feet (ten stories) high and growing each year by 10 feet (one story). The only industrial installations for treating these wastes are located in France, Great Britain, and more recently in Russia and Japan.[35]

The situation is no better in the United States. Commercial nuclear reactors have already accumulated 44,000 tonnes of depleted fuel on-site, and each year the one-hundred operating reactors produce about 2,000 tonnes of depleted fuel.

Although many supporters of nuclear power dismiss the problem of wastes or discuss it simplistically, these numbers do not allow us to underestimate it. No country in the world has yet managed to develop and place into operation a geological storage site for highly radioactive waste from civilian sources.

The subject deeply divides scientists and experts. Some are convinced that geological storage is possible without serious risks. Others maintain that it is extremely risky to store significant amounts of such highly radioactive material in the subsoil. Still others believe that it is possible to place the most dangerous wastes in suitable bunkers until a more appropriate way of treating them can be developed, when better technology comes along.

After decades of discussion, some countries have only managed to identify sites where they could build deposits for the most dangerous nuclear wastes. The start date for the work keeps slipping because of uncertainty about the amount of radioactive material that can actually be stored at the sites or other problems. For example, the United States identified a site at Yucca Mountain, Nevada, which initially could accommodate all the radioactive waste produced so far in the country. While discussions continue about reducing the storage capacity of the site to protect the environment, the start date has been slipping. Just a few years ago, the date was set at 2017, which recently became 2020, until the new U.S. administration declared that the site will not open at all, and that all related work is going to stop. Whether or not it will be re-initiated later is unknown, but certainly, 2020 is no longer feasible. Meanwhile, the United States has already spent more than $7 billion on studies and feasibility projects to identify storage sites.[36]

There is also great uncertainty about decommissioning costs, aggravated by a lack of transparency and repeated underestimation of costs by the businesses concerned. To be fair, the specific costs of the various projects are driven by the safety and environmental regulations of the individual countries, the size of the power plants, the technology they used, and the actual level of pollution produced at the site.

Under the best conditions, decommissioning a 1,000-MW, water-cooled power plant could cost on average $500 million. This could go up to $2.6 billion for some English gas-cooled reactors of the Magnox type, which is more or less what it would cost to build a new nuclear power plant of the same size.[37] Decommissioning costs for a nuclear power plant should be calculated when the plant is built, and payments into a decommissioning fund should be part of users' electricity bills throughout the life of the plant.

During the 1970s and '80s, many power plant builders underestimated the costs of decommissioning and handling radioactive waste, putting off the problems until the reactors would actually be shut down. It is also true that over time environmental and safety regulations increased a little bit everywhere, becoming more realistic and raising the cost of decommissioning. However, the initial and repeated underestimation of expenses was so enormous that it threatened the financial viability of entire countries. This happened when some countries like Great Britain and France were privatizing public companies in the nuclear power sector. For example, in 2001, British Nuclear Fuels Ltd. (a state-owned company) accumulated

£35 billion in costs, mainly to decommission its nuclear reprocessing installations.

A sort of Judgment Day is looming for many power plants built in the 1970s and '80s. Operational extensions of ten or twenty years will be granted to many power plants, but the oldest and most obsolete ones will still need to be shut down and dismantled. The public will have to pay, another heavy blow to the credibility of nuclear power.

Finally, the production of atomic energy carries with it the constant risk of proliferating weapons of mass destruction. This can happen through the illegal transfer of technology for enriching uranium and reprocessing the used fuel, and by the illegal sale of fissionable and other materials with lower radioactivity, which could be used to make radioactive ("dirty") conventional bombs.

The Future

It is difficult for me to imagine solving our future energy and environmental problems while reducing the role of nuclear power. It is unrealistic to think that renewable sources can compensate for a decline in nuclear power over the next twenty years, given the technical problems that still limit their ability to generate sufficient electricity at acceptable cost.

Without nuclear power, the carbon dioxide toll that worries us would be worse than it is today. The same goes for the next two decades.

True, there are optimists about the medium- to long-term prospects of nuclear energy. In 2008, the IAEA boosted its projection of nuclear generating capacity for 2020 from 437–542 GW to 473–748 GW, up to twice the capacity in place today.[38]

These estimates seem overly optimistic, like the predictions for total power-generating capacity expected in the world. The latter could grow within a band of 1,000–1,600 GW by 2030, while the increase in nuclear capacity would be 100–375 GW. Even allowing an increase in the middle of this range, nuclear power would still not recover its position or demand compared to fossil fuels. Even the usually optimistic World Nuclear Association, which includes the most important nuclear power producers of the world, admitted that the net growth of nuclear power electrical generation is unlikely to keep up with soaring electricity demand by 2030.[39]

I believe that time is working against a real renaissance of nuclear power, at least in the next twenty years. Only Asia will see an increase in this source of energy. Elsewhere, especially in the industrialized countries, it will be unlikely to bloom again on a large scale. Perhaps during the next two decades, a new debate will emerge about the rebirth of nuclear power, rather than a real rebirth. This period of incubation will yield results only if serious and responsible dialogue takes place on the pros and cons of atomic energy and finding a solution to the problems that still surround it.

I do not think that we will move public opinion on this subject with arguments such as "Others have it" or "We are taking the risks without receiving the advantages" or dismissing concerns about the overall cost and safety of the plants and what to do with the radioactive waste. Nor will we be able to change public opinion by insisting on having secure national energy supplies, which nuclear power could provide more assuredly than fossil fuels. Nuclear power will only be able to replace crude oil for limited purposes such as generating electricity. It will never replace petroleum for transportation. However, it could replace or compensate the growing demand for natural gas or coal in many countries.

Then there is the issue of the safety of the power plants themselves. Passive safety designs can even eliminate the risks from a failure of the reactor core, where the nuclear reaction takes place. Passive safety means that, in the event of an accident, the plant goes into a fail-safe state because its design relies on physics, not on intervention systems that require sensors, motors, or human operators. For example, should the pressure vessel of a small EPR reactor break, the core would melt at a temperature of about 2,000°C and be recovered, confined, and cooled at the base of the building. The melted material would spread by gravity over an inclined floor of refractory material, thinning out. At the same time, water from a nearby pool would cool it so that it would gradually solidify. Nothing dangerous should escape.

Unfortunately, although they have already been built on paper, passive security plants have yet to prove their effectiveness in any industrial application in the world. This does not play in favor of a new impulse for nuclear power.

It would be better to explain calmly that, as with any human endeavor, the risk of an accident can never be completely eliminated, but it can be reduced to such low levels as to be considered remote and manageable and therefore with no impact on public health. Historical evidence shows that the victims of unsafe nuclear power plants have been infinitesimally fewer than those of any other type of energy system.

Unfortunately, the extreme delay in developing geological storage sites for highly radioactive waste will continue to keep the brakes on any resurgence of nuclear power. Nor do I see political acceptance for building more closed-cycle plants, which can recycle most of the waste. On this subject, I share the concerns of MIT, which sees the closed-cycle power plant as a source of plutonium that could be diverted illegally to terrorist countries or groups.

Finally, the decommissioning of power plants remains an open issue. The problem is less pressing for new-generation plants, spreading out the cost of decommissioning over sixty years of electrical bills. Of course, governments must enforce strict legislation to ensure that those funds are actually set aside for their final purpose.

Perhaps one solution that could accelerate a new age for nuclear power would be to focus on smaller plants (200–300 MW). Even today, we can obtain excellent levels of passive safety in these plants, as well as many other advantages from the longer life of the power plants themselves. Some plants have been designed with easily replaceable modules. Of course, we could not relaunch this source of energy without offering local communities some sort of material advantages that will encourage them to host these power plants on their territory. There should be direct and indirect incentives, capable of activating mechanisms that promote improved quality of life and development. One path might be to distribute electricity at its wholesale production cost (tax-free) to local populations and businesses in the area. This could be done so as to have areas for potential sites compete with each other, with a fixed deadline for applications. This road does not lack problems, but I cannot see any other real possibilities for reducing the resistance to nuclear power in countries where local sentiment outweighs all other authority.

What we can expect for sure is that a declining role for nuclear energy will add to the already daunting challenge facing our world: reducing greenhouse gas emissions and pollution, while striving to produce more energy.

Statistics on Nuclear Power

Table 5.1. Proven uranium reserves (Top 10 countries and world total, 2008)

Country	*Proven reserves (tonnes of uranium extractable at a cost of less than $40/kg)*
Australia	709,000
Canada	270,100
Kazakhstan	235,500
Brazil	139,600
South Africa	114,900
Namibia	56,000
Uzbekistan	55,200
Russia	47,500
Jordan	44,000
China	31,800
World total	**1,766,400**

Table 5.2. Uranium production (Top 10 countries and world total, 2008)

Country	*Production (tonnes)*
Canada	9,000
Kazakhstan	8,521
Australia	8,430
Namibia	4,366
Russia	3,521
Niger	3,032
Uzbekistan	2,338
United States	1,430
Ukraine	800
China	769
World total	**43,853**

Table 5.3. Uranium consumption (Top 10 countries and world total, 2008)

Country	*Consumption (tonnes)*
United States	22,825
France	9,000
Japan	8,790
Russia	4,100
Germany	3,490
South Korea	3,200
Ukraine	2,480
Canada	1,900
United Kingdom	1,900
Sweden	1,600
World total	**69,110**

CHAPTER 6

The Weight of Water

In addition to sustaining our life each day, water has provided us with energy since ancient times. At first, it provided only mechanical energy. From the middle of the nineteenth century, it was able to provide electrical energy, thanks to the invention of the hydraulic turbine. Hydroelectric power spread throughout the world, making water one of the most important sources of electricity for a good part of the twentieth century.

Today water could represent one of the energy hopes for our future. By itself, it accounts for a little more than 2 percent of the primary energy and about 16 percent of the electricity consumed in the world.[1] Significantly, in sixty-five countries, hydropower still produces more than half of the electricity.

It is clean energy and theoretically renewable, with generally competitive costs. Hydroelectric plants are absolutely the longest-lasting structures that produce electricity. Some plants can respond on very short notice to peaks in demand. Yet, this promising façade hides many problems.

Yes, hydroelectric power is renewable, as long as there is no shortage of rainfall and a drought does not turn water into an exhaustible resource. Historical costs in general appear competitive, but in many cases, the assessments were not carried out correctly. To provide large quantities of electrical power and be able to control the output, it is necessary to build dams. These can have a negative impact on the environment, the land, and the people who live there. We must also not forget that the vital, priority use of water is to satisfy the needs of people, agriculture, and industry. These needs can sometimes conflict with the production of electricity.

Finally, exploiting the water in rivers for electrical purposes is not always possible, especially when those rivers cross hundreds of miles of thick forests or deserts far from inhabited centers. This is the situation in Africa or Latin America.

Thus, although the hydroelectric production potential that is theoretically available and economically exploitable in some areas of the planet is still very high, what can actually be used is much less. To this bitter reality, we must add the environmental and social challenges that have further undermined the feasibility of increasing hydroelectric generation, especially since the end of the last century. However, without a substantial contribution of electricity from hydropower, the energy picture of the future is destined to be much more complicated.

The Rise of an Ancient Resource and Its Problems

For much of the twentieth century, hydroelectric power represented the principal source of electric power in the world, dominating the scene in many countries. For example, in 1960, it still supplied 84 percent of Italian electrical requirements, 51 percent of Japanese demand, and almost all of the electrical needs of Africa and less-developed areas of the world.

Credit for the hydraulic turbine goes to French engineer Benoît Fourneyron. In 1827, he developed the first modern prototype capable of exploiting the energy in water to generate electricity. It took a long time for the first commercial hydroelectric plant to come on line in 1882 in Wisconsin. That plant is still operating today.

Another important step in the development of the sector was taken in 1901, when George Westinghouse installed an alternating-current transmission system at the hydroelectric plant at Niagara Falls between New York and Ontario. This was also a turning point in the so-called Current War during the early days of electricity production.

Westinghouse had been sparring with Thomas A. Edison about electrical transmission systems. Edison espoused the cause of direct current (DC), for which he had built his first power generation and distribution systems. Westinghouse favored alternating current (AC). Initially, the prevailing use for electricity was public lighting, which used 110 volts, and this became the standard for the lines carrying the current. As the use of electricity spread, the limits of direct current became evident. The loss of electricity in transmission (*line loss*) increases with the length of the transmission line; line loss is greater at lower voltages. Unlike DC, alternating-current voltage could be higher during transmission and then lowered through transformers at destination for final use. AC won the war.

This important development made it possible for users to take advantage of electricity from distant water sources. Turbines and dams grew exponentially in size as hydroelectric power expanded rapidly. The center of activity was in the United States, where dam construction peaked in the 1930s and '40s. After World War II, new construction was concentrated in developing countries. Of the 45,000 large dams in existence today, about 40,000 have been built since 1950, the vast majority of them in developing countries. Half were built in China alone.[2]

The criteria for building dams during the postwar rush left much to be desired. Engineering errors, poor building materials, inadequate safety standards, and lack of flood resistance caused many dramatic accidents, which in some parts of the world caused thousands of casualties.

Among others, scientists and fishermen joined the growing throng of critics of these imposing projects. They blamed the dams for modifying the hydrologic balance of the areas involved and the habitats of fish and other aquatic organisms, causing their disappearance from the artificially modified watercourses.

The most vociferous protests emerged gradually, as evidence multiplied of the violent and sometimes brutal intimidation used by the governments of developing countries to force entire populations to abandon villages and cities scheduled to be flooded. During the 1980s, the growing opposition finally embarrassed the World Bank, which over the previous decade had guaranteed many of the loans needed to develop large hydroelectric installations in developing countries. By the 1990s, the flow of accusations of recurring brutality associated with the construction of large dams swelled like a river in full flood, shining an international spotlight on previously ignored incidents.

The most tragic of these was the massacre of almost four hundred Maya Achí Indians in the village of Río Negro, Guatemala, between 1980 and 1982. Paramilitary bands and Guatemalan soldiers savagely kidnapped, tortured, and killed the men, women, and children of the village, which had opposed the construction of the dam at Chixoy, badly wanted by the dictatorial regime of the country and financed by the World Bank. Even after the massacres, the bank provided additional incentives for the project.[3]

After human rights activists exposed this and other cover-ups, the World Bank and the World Conservation Union established the World Commission on Dams. They charged it with carrying out a broad analysis and assessment of the state of dams throughout the world, establishing the truth about the cost, environmental impact, and social implications associated with their construction.[4] The commission worked for two years, producing the most extensive and complete report ever prepared on the subject. It tore the veil from the many simplifications that had accompanied dam development.

In brief the report revealed that builders of large dams had inflated the estimates of economic and social benefits of the completed projects. The benefits always proved more modest when carefully analyzed, while the social and environmental impact of dam construction had generally been ignored or greatly underestimated.

In many developing countries, powerful pro-dam lobbies had influenced local governments, while opposition to dam construction was badly tolerated or repressed. Many of the public works already built exhibited precarious safety conditions, particularly in developing countries. The report underscored that large dams could also constitute a significant source of greenhouse gas emissions from the decomposition of vegetable detritus collecting in their reservoirs.

The commission report marked a fundamental break in the vision and concept of dams, reinforcing a rethinking process that was already under way in many advanced countries. For example, by the 1990s, the dismantling of large dams had already exceeded the construction of new ones in the United States.

Naturally, dams remained a major opportunity, in some cases a vital chance, for development in vast areas of our planet, ensuring survival of entire populations and the agricultural development of regions that were otherwise condemned to permanent drought. However, the approach to building dams had to be completely rethought, with a serious risk analysis of each project, as well as adequate assessment of the social and environmental impact. These principles have rapidly caught on in the developed countries, but they are still not being applied in many poor countries.

Meanwhile, the golden age of hydroelectric power is over. The increase in hydroelectric generating power was never able to keep up with the constant increase in energy demand. In 1995, hydroelectric generation represented almost 19 percent of world electrical production, but by 2005, it had dropped to 16 percent.[5]

The State of the Art

Hydroelectric power comes from the energy in the flows and falls of watercourses, which moves turbines connected to electric generators.

Small hydroelectric plants (the most common in the world) simply take advantage of the flow between two points of a river that is flowing downhill (called *local relief*). On average, the difference in elevation is about 30–50 feet (10–15 meters). Because the local relief is relatively small, the turbines must be able to count on an adequate water flow. Classified as *flowing-water plants*, these systems do not require the construction of dams (except as needed to guide the water) nor reservoirs to accumulate the water.

In addition to their relatively small size, flowing-water plants are limited to producing electrical power according to the rhythms of nature (the flow of available water), which does not always coincide with the rhythms of electrical consumption. They can serve as basic production installations, but they cannot respond to sudden peaks in demand.

To take full advantage of the potential of water, it must be accumulated in artificial reservoirs and allowed to flow through specially built conduits (*penstocks*) to the turbines of the generators. Dams are needed to make this happen.

The difference in the capacity of the two types of hydroelectric systems is stunning. The largest flowing-water plants can supply a few megawatts of capacity. Dams include some of the biggest electrical generation systems in the world, with installed capacities much greater than 1 GW.[6]

The biggest operating dam on the planet is at Itaipú, between Paraguay and Brazil. It has a capacity of more than 13 GW, but it will soon be surpassed by the Three Gorges Dam in China, with an expected operating capacity of more than 18 GW.[7] In any case, the six largest dams in the world are connected to electrical generation capacity of at least 6.4 GW each. Only one other facility, the

Japanese nuclear power station of Kashiwazaki-Kariwa (8.2 GW), is in that league.[8]

Dam systems offer other technical and economic benefits. The initial investment can vary between $500 and $2,000 per installed kilowatt for installations greater than 250 MW. Naturally, the larger the capacity, the lower the unit cost, but the greater the timelines and the problems of the project. However, the costs of smaller systems that do not use dams or basins are much higher, between $2,000 and $4,000 per installed kilowatt for installations under 20 kilowatts.

Although the initial outlay is rather high, the investment in large dams is compensated by the long life of the infrastructure, the longest lasting of all energy-producing structures. Some dams have been operating for more than a century. In addition, operating costs are very low, because once the initial production expenses are depreciated, the operating costs are very competitive compared to other energy sources. They can also respond quickly to sudden peaks in demand, opening sluices and bringing turbines on line as needed.

There is a special category of hydroelectric systems, the *pumped hydroelectric storage plant,* which is really a type of artificial reservoir system. It consists of two water collection reservoirs, one higher than the other. When demand is low, typically at night, electric pumps move the water in the lower basin to the upper basin, from which it flows to produce electricity when demand increases.

Pumped hydroelectric storage systems worldwide have a capacity of more than 90 GW. More than 40 percent of this is in Europe, almost 5 percent of it in Italy, which has been using this type of installation since the late nineteenth century.

Naturally, pumped hydroelectric storage systems are more expensive than dams, and they require more electrical energy to operate than they produce. In other words, they have negative energy balance (with a typical yield between 70 percent and 80 percent of the energy input that they require). Nevertheless, they have the great advantage of storing the potential energy of the water in the upper reservoir.

Some support associating pumped hydroelectric storage systems with renewable energy installations, such as wind turbines or solar panels. Using the electricity from the renewable sources during periods of low demand to pump water might yield an ideal marriage of sun, wind, and water. The idea may prove very expensive and not always feasible, but it is interesting enough to warrant further study.

In 2007, the World Energy Council (WEC) provided an estimate of the theoretical, technical, and economic hydroelectric potential available worldwide.[9] It should be noted up front that calculations of this type are useful for obtaining an idea of the orders of magnitude involved, but the data should be considered very carefully.

The *theoretical potential* comes from an estimate of the volumetric flow of water of all the watercourses in the world and the corresponding assessment of the electrical energy that could be obtained from them, assuming a conversion efficiency of 100 percent. This potential is estimated to be more than 41,000 terawatt-hours (TWh) per year (more than twice the entire production of electricity in the world in 2005).

The *technical potential* corresponds to just that portion of the water flow that can be exploited for electricity with available technologies, disregarding the costs associated with production. The WEC estimated that the worldwide technical potential is almost 16,500 TWh per year.

The *economic potential* considers the cost of exploiting the technical potential, measuring both the specific conditions in each country and comparing the cost of hydropower to other sources of energy. This determines if hydroelectric power is in fact competitive. Considering these variables, the exploitable hydroelectric potential becomes noticeably less, about 8,800 TWh per year or a little more than half of the technical potential. Naturally, changes in technology or an increase in the cost of competing energy sources could improve the situation for hydroelectricity.

The WEC estimates are more or less shared by other sources. They point to a very significant hydroelectric potential for a world that in 2005 produced more than 18,000 TWh of electricity, only 2,900 of which were hydroelectric. In other words, only one-third of the energy available from water was used, but the hydroelectric potential could have provided half of world production.

The WEC assessments were based on early theoretical assumptions. They did not take into account important elements that seriously limit the real possibilities for developing hydroelectric power.

The first problem is the fact that a large part of the fresh water on our planet is needed for human survival and for irrigation. For that matter, most dams in the world today were built for irrigation, industrial and domestic consumption, flood control, and other uses, not for producing electricity. In Europe, only 33 percent of the dams are used for hydroelectric production (20 percent are used for irrigation). In North and Central America, only 10 percent of the dams generate electricity; in South America, 25 percent. In Asia, just 7 percent of the dams generate electricity (67 percent are used for irrigation). In Africa, 75 percent of the dams are used to provide water to the population or for irrigation.

Water consumption by other activities can conflict with electricity generation, even if the water used for electrical generation is recycled in the hydrologic circuit and thus is not lost. Water is also becoming a critical issue for a growing part of humanity. Indeed, it is the great problem of our century. Even where water exists, its continued availability is not assured.

This leads to the second problem, which is both substantial and semantic: reliability. Although hydroelectric energy is technically classified as a renewable resource (and in principle it is) and is considered reliable over time, the whims of nature or the cycles of the weather can turn that around. Just one year of low rainfall can produce shortages of electrical power, as watercourses and reservoirs lose part of their capacity. Then in other years, heavy rains can cause structural problems. A good example is Brazil, a country that by the end of the 1990s depended on water for 90 percent of its electrical requirements. After years of noticeable reductions in rainfall, the South American giant saw its hydroelectric capacity dropping just as electrical consumption in the country was rising at a high rate. This unexpected unreliability of water moved the government in Brasilia to diversify the power generation mix to natural gas, fuel oil, and nuclear power.

The third problem affecting the real amount of hydroelectric potential that can be exploited concerns the construction of large installations. Developing hydroelectric capacity on a vast scale requires modifying the political, human, and environmental geography of the land, building dams that displace thousands of persons, eliminating entire villages, modifying the hydrologic cycle of vast areas, and reducing the fluvial fauna. Negative examples abound.

The most recent sensational case of the evil consequences of a large hydroelectric project is the Three Gorges Dam in China. Dozens of villages, towns, and archaeological sites were buried by the waters of the Yangtze River, and more than a million people were moved from their homes, which were scheduled to be flooded. In September 2007, even the Chinese authorities recognized that the dam might carry "hidden dangers" (in particular, landslides, land erosion, and pollution) sufficient to generate an "environmental disaster."[10]

Sometimes the construction of dams or other upstream exploitation of rivers that cross borders can give rise to international disputes, conflicts, or enduring cycles of recriminations. For example, the construction of a series of twenty-two dams in the area where the Tigris and the Euphrates rivers rise in Turkey (the Southeastern Anatolia Project) threatens water levels in Iraq and Syria. This has led to increasing tension between these two countries and Turkey.[11]

Hydroelectric power runs into other kinds of problems. There is no doubt that rivers like the Niger or the Congo could provide vast amounts of electrical power, but getting from theory to reality is virtually impossible. For some time, there have been dreams of a construction project on the Congo, the river with the highest hydroelectric potential in sub-Saharan Africa. It would involve a series of dams and power plants with an installed capacity of almost 40 TW. The reluctance of international investors to undertake large commitments in the region makes the project difficult enough, but moving the electricity from the power plants through hundreds of miles of forests to the principal cities makes it practically impossible.

In many developing countries, opposition by nongovernmental organizations has been growing against the construction of large dams. For example, in Brazil the construction of two new dams on the Madeira River was approved by the Brazilian Environmental Protection Agency, but still triggered boycotts organized by environmental and human rights groups.

The Future

I have tried to explain why I find it difficult to believe that increasing hydroelectric power on a global level can satisfy our growing need for energy.

One of the great challenges of this century will be to safeguard the water that we have and to make it available to the many people who do not have it at all or who are losing the water that they have. The risk of drought or desertification of significant areas of the Earth makes the proliferation of large hydroelectric projects problematic. In regions where there is significant water potential, it proves to be more theoretical than practical. The industrialized countries are close to saturating capacity.

This does not mean that the force of water cannot make an important contribution to the energy challenge that faces us. A realistic and prudent assessment of the hydroelectric potential that could be cost-effectively exploited in the next two decades reveals at least 1,500 GW of installed capacity. That is twice what is in place in the world today, and it would allow hydroelectric power to increase its share of the worldwide energy mix by a few percentage points in the near term.

Much of the additional capacity is in Asia, Africa, and Latin America and would require the construction of large dams. On this subject, some clear thinking is needed.

Recent criticism of large hydroelectric projects is certainly justified if it concerns the irresponsible indifference to environmental and human problems that characterized these projects in the past. However, there is more awareness today. A balanced judgment of these structures cannot fail to take account of their indispensable role in meeting the vital needs of vast areas of the world, which would otherwise suffer drought or the absence of electricity. The alternatives to hydroelectric dams do not paint a pretty picture: darkness or greater consumption of fossil fuels, with their load of greenhouse gases and local pollutants.

An important role could be played by smaller projects, especially distributed microgeneration projects. It is true that there are 45,000 large dams in the world, but there are also almost 300,000 small dams that nobly carry out their work without the human and environmental problems of the larger projects. On paper, we cannot deny that the large dams bring about a reduction in the cost of electricity produced, simply by economies of scale. However, a proper assessment of the real costs of displacing populations, adopting strict safety standards, building systems that contain the damage to fluvial life systems and the hydrologic cycle, and paying for maintenance and insurance against major risks might tip the economic balance back in favor of small dams.

Moving from large-scale sites to medium-size ones spread around the land, offers a less uncertain and less problematic outlook for developing hydroelectric power. This hope is reinforced by continued technological improvement in the turbines and hydroelectric systems overall. These open the way to the possibility of producing greater amounts of electrical power for the same amount of flow or water height. Technology might even increase the potential of flowing-water plants.

Water will probably not become a principal energy source in our future. However, if we reduce the share of hydroelectric energy (and nuclear power) in the world energy mix, it will be impossible to curtail the dominance of fossil fuels.

Statistics on Hydroelectric Power

Table 6.1. Installed hydroelectric generation capacity (Top 10 countries and world total, 2008)

Country	*Installed capacity (GW)*
China	165
United States	100
Brazil	78
Canada	74
Japan	48
Russia	47
India	37
Norway	29
France	26
Italy	21
Total	**947**

Table 6.2. Hydroelectric energy production (Top 10 countries and world total, 2008)

Country	*Production* (TWh)*
China	563
Canada	372
Brazil	365
United States	252
Russia	181
Norway	140
India	113
Venezuela	87
Japan	74
Sweden	69
Total	**3,191**

* Excluding pumped storage plants

CHAPTER 7

Biofuels: Behind the Smoke and Mirrors

Before dealing directly with the subject of biofuels, we need to make a short digression into the broader category of resources from which they are obtained, usually called *biomass*. Biomass yields not only biofuels but also other energy vectors such as heat and electricity. However, the term itself creates much confusion, especially when it is associated with the idea of clean energy. In reality, much of the energy produced from biomass is dirty, polluting, and harmful. This is no small matter.

After fossil fuels, biomass is the most important source of energy for humanity, supplying about 10 percent of the primary energy consumption in the world. Only 40 percent of biomass is used in a modern way, and it covers only 4 percent of the worldwide demand for primary energy, while the better-known hydropower supplies just 2 percent. The other 60 percent is far from the public image conveyed by the prefix *bio-*, that is, something clean and environmentally friendly. In reality, most biomass includes wood, animal and vegetable leftovers, and even dried dung, burned for cooking, heating, and light. This is no more and no less than what our Neolithic ancestors used, with damaging consequences for both the environment and the health of the people using it.

China, India, and the countries of sub-Saharan Africa are the principal consumers of biomass, but worldwide more than 2.5 billion people draw the energy they need to cook and heat their dwellings from animal and vegetable garbage. The result is that deaths caused by using these kinds of fuels inside homes outnumber deaths from malaria, while the deforestation involved has had a devastating impact on the environment.

Specifically because of the dangerous health consequences and low efficiency of these sources, the United Nations Millennium Project proposes cutting their use in half by 2015, introducing more evolved systems of energy conversion and modern fuels.[1] There are also consequences for climate change: burning biomass emits *black carbon* (a soot),[2] considered by scientists the second most important contributor to rising global temperatures after carbon dioxide.[3]

Biomass can definitely be a more versatile energy source than other renewables. It can be stored and used on-site for heating (more than 60 percent of the modern use of biomass), for power generation (about 15 percent, mostly in combination with heat), and for biofuels (only 5 percent).[4] Currently, biofuels are the most versatile products derived from biomass, as is already obvious in the United States. In the first place, the technology for generating heat and electricity from biomass is mature, although there is still untapped potential for these uses of biomass. Furthermore, biofuels are the only renewable source that can be directly used as a substitute for oil in the transportation sector. That makes them new and exciting. Thus, both worldwide attention and many ambitious government policies have focused on biofuels,[5] and this chapter will concentrate on this aspect. I will try to explain the possibilities and the doubts about biofuels.

A Real Competitor of Gasoline and Diesel Fuel?

The term *biofuels* indicates compounds derived from the transformation of renewable raw materials such as cereals, plants, vegetable oils, sugary crops, and other organic substances like wood scraps and animal fats.

Different fuels are obtained from different biomasses. First generation *bioethanol,* a direct competitor of gasoline, comes from fermenting the starch or sugar in some sugar-rich or cereal crops (e.g., sugar beets, sugarcane, corn, barley, and other cereals).[6] First-generation *biodiesel,* a direct competitor of diesel fuel, is produced from vegetable oils extracted from fruits or seeds of specific plants or crops (sunflower, rape, palm, etc.), through a transesterification process.[7]

Advanced biofuels, the second-generation biofuels, are derived from agricultural and forest residues (waste from cereal crops, timber operations, pruning, etc.), and from nonfood crops. Woody-cellulosic materials can be turned into bioethanol through a biochemical process (hydrolysis and successive fermentation of sugars)[8] or transformed into biodiesel through a thermochemical process (*biomass-to-liquids*).[9]

It is worth noting that some confusion exists about the exact boundaries of first- and second-generation biofuels. The broad distinction refers to the processes used to convert them from biomass. However, most experts also attribute other features to second-generation biofuels, mainly nonfood competition (i.e., they are produced from crops that do not enter the human food chain) and a better environment and climate footprint (reduction of pollution and greenhouse gas emissions).

For example, Brazilian ethanol produced from sugarcane belongs to the first generation because of its production process. Yet its carbon footprint is better than most second-generation biofuels, because on a lifecycle basis its combustion produces 60 percent less carbon dioxide than gasoline. Similarly, biodiesel produced from some purpose-grown energy crops, such as *Jatropha curcas* (which does not compete with food production), is generally considered a second-generation biofuel, even though it is obtained through a first-generation process. Although second-generation biofuels have not achieved commercial viability yet, most agree that these advanced biofuels could provide effective answers to many of the concerns entangling the first-generation biofuels.

Biofuels can be used in existing engines by adding them to gasoline (up to 10 percent) or diesel fuel (up to 20 percent) without damaging the engine. However, the 10 percent bioethanol blend in old vehicles can corrode the internal surface of the fuel rails. This happened in the United States, where Toyota recalled 214,500 Lexus cars because of problems with bioethanol. Despite that, the U.S. ethanol industry is pushing the Environmental Protection Agency to increase the amount of bioethanol blended into gasoline from the current maximum of 10 percent (E10) to as much as 15 percent (E15), even though only 3 percent of U.S. motor vehicles are designed to run on fuel containing more than 10 percent of bioethanol.

There are also "flex-fuel" (flexible fuel) vehicles that can use bioethanol and gasoline in any proportion, including 100 percent bioethanol; so far, however, these vehicles operate only in Brazil and, to a lesser extent, in the United States and Sweden. On the other hand, 100 percent biodiesel has been available for several years in Germany, but it can only be used in specially modified diesel engines.

The first problem with these motor fuels concerns the problem of quantity, that is, the amount of energy they really provide. In 2007, the two largest providers of bioethanol in the world, the United States and Brazil, produced about 90 percent of it, the equivalent of almost 470,000 barrels of petroleum per day. Meanwhile, the world consumed almost 86 million barrels of petroleum per day (in terms of energy content). Put another way, global production of bioethanol in energy terms was equal to 34 billion liters (9 billion gallons) of "biological" gasoline, compared to a worldwide gasoline consumption of about 1.2 trillion liters (315 billion gallons), or just 3 percent. This was in spite of the fact that the biofuel industry in the United States and Brazil has been heavily subsidized by the government since it took off in the middle of the 1970s.

Unlike bioethanol, biodiesel is marketed today mainly in Europe, where about 60 percent of worldwide production is concentrated. Its numbers are even more limited:[10] a total of 5.6 billion liters of diesel-equivalent against total European diesel consumption of about 200 billion liters—again, 3 percent. On a world level, 9.4 billion liters of diesel-equivalent were consumed, compared to the worldwide diesel consumption of about 770 billions, or 1.2 percent.

Such modest numbers can be explained by the fact that the amount of energy produced by biofuels is very low for each acre cultivated. This means that to obtain barely noticeable amounts of biofuel, vast expanses of land are needed. This is the curse of many alternatives to fossil fuels and nuclear power, and it affects biofuels, too. Their energy density is very low, and their power density is equally meager. A few examples can help illustrate the problem.

The United States is by far the top producer of corn in the world, growing about 40 percent of global supply. It is also the biggest exporter, with American corn accounting for almost 60 percent of world corn exports.[11] Currently, almost 30 percent of the U.S. corn supply is used for ethanol production. *If all the remaining corn production were to be turned into biofuels, it would replace only about 5 percent of U.S. petroleum demand.*[12] This does not even consider the devastating effects of such a transformation in terms of the cost, environmental damage, and disruption of food supplies.

The result is no different moving from a big country like the United States to a smaller one like Italy. The total area of my country is 30 million hectares (75 million acres), of which only about 13 million hectares are arable. Converting all arable land to produce rapeseed oil would supply only about 10 percent of the Italian petroleum requirement.[13] That would be a terrible devastation of the land for an insignificant result. Removing wheat fields, rice paddies, olive groves, vineyards, and so on would eliminate 100 percent of the world supply of authentic Italian pasta, risotto, balsamic vinegar, extra-virgin olive oil, and wine.

When speaking about the productivity of traditional agriculture in terms of biofuels, one must always pay attention to two different aspects, which are often confused. The first is the productivity of a crop selected for a given territory, that is, how many harvests of seeds or fruits per year that plant can guarantee. The second is the amount of sugars, starches, or vegetable oils (the raw materials of bioethanol and biodiesel) that can be extracted from those harvests.

The productivity of different crops varies greatly, not only because of their own characteristics but also because of environmental conditions (soil, water balance in the area, sun, climate, etc.). In parts of Asia, such as Indonesia and Malaysia, palm trees can yield 5–6 tonnes of oil per hectare (about 2.5 tons per acre) per year for making biodiesel, but to the detriment of the environment, because the palm tree flourishes in the elevated humidity of deforested areas. In sub-Saharan Africa, the oil yield is much lower, about 2.5 tonnes per hectare, without irrigation.

Unlike palm trees, which produce almost continuously, grassy crops like soy and rape yield only one harvest per year. For this reason, although the seeds of some grassy crops have a high oil content, the annual productivity of vegetable oil per hectare is much reduced (0.5–1.0 tonnes of oil per hectare). The same goes for crops used to produce bioethanol.

Sugar beets and sugarcane yield abundant harvests, usually more than 50 tonnes per hectare, up to 100 tonnes in ideal areas. However, they have a very limited sugar content (about 15–18 percent), which is the raw material for ethanol. On the other hand, American producers using corn have smaller harvests (6–9 tonnes per hectare, or 2.7–4.0 tons per acre), but the amount of starch in the kernels that can be turned into sugars is about 30 percent. In the end, crops are either highly productive with a low energy yield or vice versa.

Depending on the place where they are planted, the results can vary significantly.[14]

Considering the entire production chain, each hectare cultivated in Europe yields 3–5 tonnes of bioethanol under ideal agricultural conditions, or 1.0–1.5 tonnes of biodiesel. Other areas of the world achieve greater productivity: 7 tonnes of bioethanol per hectare of sugarcane in Brazil and 4–5 tonnes of biodiesel from palm oil in Malaysia (but with the severe environmental impact just mentioned).

Considering these limits, the only global option for producing significant amounts of biofuels is to turn to the areas of the planet that are best suited for it, in particular Latin America and Africa.[15] However, we should not harbor any illusions about the macroscopic impact of this type of initiative, because the basic problem remains: The amount of energy that biofuels from traditional crops can supply is limited when compared to the large numbers of world energy demand. The press routinely reports statements from self-styled experts and even scientists with advanced degrees, debating the revolutionary virtues of plants with exotic names. In reality, no known plant has overcome the problem that any crop has only modest energy content.

The second big problem with biofuels is their cost. With oil prices in the range of $60–$70 per barrel (as in 2006–2007), a liter of gasoline costs on average 50 cents. It costs about 30 cents to produce an amount of ethanol holding the same energy content as a liter of gasoline in Brazil, the only country where biofuels come close to really competing with gasoline. In fact, the cost of bioethanol increases to an average of 75 cents per gasoline-equivalent liter ($2.85 per gallon) in the United States and about $1.30 per liter in Europe.

Many people make the error of equating a liter of gasoline with a liter of bioethanol. Bioethanol contains oxygen and therefore has a lower heating value or chemical energy content, 30 percent less than that of gasoline. As a result, a tank of bioethanol will not go as far as a tank of gasoline. Furthermore, transportation, blending, and distribution can add another 20 cents per liter (75 cents per gallon) at the pump.

The figures for biodiesel are no better. Although its heat content is only a little less than that of petroleum-derived diesel (11 percent less), it costs more to produce than bioethanol, which cancels out its energy advantage.

As I have mentioned, the cost of producing the different biofuels can vary greatly, but this does not change the basic reality that the most efficient production methods for first-generation biofuels can stand on their own (without government subsidy) only when the cost of petroleum goes over $70 per barrel. In the case of biodiesel produced from rapeseed oil in Europe, an MIT study estimated the breakeven oil price to run as high as $160 per barrel.[16]

Only Brazil has managed to provide competitive bioethanol, and for this reason, the country is the subject of great international interest. However, *the case of Brazil remains an exception.* In fact, its success is due to peculiar circumstances that cannot be replicated except in very limited regions of the planet. First, there is the availability of immense areas of arable land (both absolute and per capita) perfectly suited for sugarcane, the crop that is the most productive for bioethanol and that consumes the least energy to obtain it.[17] Second, water and low-cost labor are widely available. Finally, the climate and temperature of the country are ideal for sugarcane growth.

For these reasons, Brazil is to biofuels as Denmark is to wind power, Saudi Arabia is to oil, and the United States is to coal. Nature has endowed these countries with specific resources and features that cannot be replicated.

Nevertheless, Brazil has not achieved these results without suffering and problems. Production of bioethanol in the South American country began in the mid-1970s with the government ProÁlcool program, which was the response of the Brazilian government to the first oil crisis. Thanks to incentives and subsidies, by the mid-1980s, more than 90 percent of the automobiles sold in Brazil were running on bioethanol.

Then, at the peak of its success, sales of bioethanol-burning vehicles collapsed. Bioethanol production dropped drastically for two reasons. The first was the crash of oil prices in 1986, which made ethanol more expensive than gasoline. The second was the rising price of sugar, which drove the sugar growers to transfer their production to exports instead of bioethanol synthesis.

Interest in the biological fuel revived only at the beginning of this century, because (again) of oil price increases, but also because of the introduction of the first flex-fuel automobiles and an increase in the price of ethanol, sustained by the growing world demand for fuel oxygenates.[18]

These market factors revitalized investment in new sugarcane plantations and biorefineries. By 2007, flex-fuel vehicles accounted for 86 percent of the vehicles sold in Brazil, at a cost comparable to gasoline-powered vehicles even without the subsidies and incentives, which had been eliminated in the early 1990s.

The Brazilian exception is not enough to herald the dawn of a new and serious competitor for gasoline and diesel. As if to illustrate the point, after a period of massive investments in ethanol refineries in the United States under the Energy Independence and Security Act of 2007 (EISA), oil prices fell sharply, and so did demand for ethanol (and biofuels in general). Those companies that took on heavy debt to increase capacity are now facing very hard times. Some have had to stop operations and file for bankruptcy. As a result, production in 2008 fell short of the EISA goal of 9 billion gallons of ethanol.

Things are not looking better for biodiesel producers in the United States, although a $1 per gallon federal tax credit has helped. However, that credit was set to expire at the end of 2009, raising concerns among operators.

The market outlook for biofuels in general is more optimistic for the medium term, as the new regulatory framework goes into effect in Europe with ambitious targets of 10 percent biofuels in the mix by 2020, and as the stimulus package of the new U.S. administration begins to have effect.

Other open questions remain for biofuels.

Are They Really More Sustainable?

The problems with biofuels do not end with the problems of energy density and cost. One of the most critical questions concerns their real environmental and energy impact.

The undoubted merit of biofuels is that they emit no sulfur oxides, polycyclic aromatic hydrocarbons, toxic substances, or carcinogens. Moreover, they have higher octane ratings (bioethanol) and cetane numbers (biodiesel) than traditional gasoline or diesel, so they can improve engine performance.

The impact of biofuels in reducing greenhouse gases is not certain, however, and their production cycle could have damaging consequences for the environment. Biofuels are often considered neutral in terms of carbon dioxide emissions, because what is emitted during their use has been previously captured by the plant during its growth. However, this underestimates the impact of their complete production cycle.[19] Careless use of available land could actually increase the overall emission of greenhouse gases. For example, it would not be smart to convert land previously forested or used for crops that absorbed more CO_2 than the biofuel crops. Furthermore, the intensive use of nitrogen fertilizers for the biofuel crops can release nitrous oxide into the atmosphere (a greenhouse gas more than three hundred times more damaging than carbon dioxide in terms of global warming).

In some countries, especially Malaysia and Indonesia, the production of vegetable oil on an industrial scale (currently intended for the food industry) has brought about massive deforestation, which could be aggravated by the production of biofuels. Trading in rain forest for palm oil plantations has a massive negative environmental impact.

Intensive and extensive cultivation of biomass also requires water—a lot of water, at least in the case of traditional crops.[20] I am amazed that so little is mentioned about this aspect, when the United Nations has designated water as the most critical problem of our century.

The environmental problems of growing crops for biofuels on a vast scale do not end here. There is also land erosion and soil depletion (already the cause of much concern in the United States); pollution of groundwater, lakes, and coastal waters from the widespread use of fertilizers and pesticides; and threats to biodiversity. Moreover, most studies so far have not addressed the environmental, energy, and economic implications of the logistics associated with transporting the biomass to the conversion plants or distributing the final product, the biofuels at the pump.

Today, the supply chain of biorefineries is based mainly on truck transport. Unlike coal slurry or oil, pipeline transport for biomass is usually not effective because biomass tends to absorb the carrier fluid (water or oil). Because of the low biomass energy density (compared to petroleum or coal), one can imagine a significant percentage of biofuel production being cannibalized to fuel its own truck fleet. This would undermine the energy and environmental balance of biofuels. In addition, bioethanol needs a dedicated supply chain to the pump because of its high volatility and its tendency to absorb water.

Although there are no definitive results on the environmental merits of biofuels, there have been some related studies. The first was conducted by David Pimentel and Tadeusz W. Patzek and published in *Natural Resources Research*. It challenged the results of previous work and asserted that no production (corn/bioethanol or soy-sunflower/biodiesel) could be considered environmentally advantageous, if one factors in the vast quantities of insecticides and herbicides that pollute the groundwater during cultivation. Producing 1 liter of bioethanol creates 13 liters of wastewater.[21] Nevertheless, a careful analysis of the study confirms that the evidence is not unequivocal, because some of the assumptions are not universally shared.[22]

The second study was published in *Science* in January 2006. It summarized the most recent analyses of the environmental balance of bioethanol and reiterated that the only process for producing it that could significantly reduce the emission of greenhouse gases was the one based on woody-cellulosic biomass.[23] In other cases, using bioethanol instead of gasoline could reduce emissions by as much as 30 percent or increase them by as much as 20 percent over the entire life cycle, depending on different variables.

The third study was done in 2007 by the Organization for Economic Cooperation and Development (OECD), with the provocative title *Biofuels: Is the Cure Worse than the Disease?* It found that, compared to traditional fuels, biofuels emit less greenhouse gas but overall cause greater damage to the environment.[24] Reducing greenhouse gas emissions from biofuels would be much more expensive than other greenhouse gas abatement programs, such as reforestation.

Certainly, the environmental sustainability of biofuels production is a crucial point, even more so if growth in the sector depends on opening it to international commerce. To ensure the quality of biofuels marketed in distant countries, a credible supranational system for certifying environmental sustainability would be needed. This is no small task.

It also takes a lot of energy to produce biofuels. The many studies done on this subject still leave more questions than answers, but it seems certain that in the best of cases the energy needed to obtain a liter of bioethanol from corn is about the same as the energy contained in the biofuel itself, except for Brazilian bioethanol. The situation seems rosier for biodiesel, the processing of which also yields secondary products (fuels like naphtha or propane, when using the most modern technologies), which help improve the energy balance significantly.

Uncertainty about the cost-effectiveness of biofuels is also fed by the wide variability of the many parameters that determine the specific production of a biofuel in a specific geographic area, with a particular climate, and under the energy yield conditions that are possible there.[25] The least that we can learn from this brief overview is that we must always avoid generalizations about biofuels and instead concentrate on a focused analysis of the individual biomass that is to be evaluated in the specific geographic context where it is being produced.

Food or Energy?

I have left for last one of the most critical and potentially destabilizing aspects of biofuels: the impact that their increased production could have on the cost of foods needed by people and livestock.[26] This issue raises some very disquieting scenarios.[27]

Without a doubt, the extensive diversion of agricultural crops intended today for human or animal nourishment into biofuels would drastically raise the price of many basic foods for human beings. This would hit the poorest populations especially hard. Similarly, an increase in the cost of feedstock for animals would increase the cost of many of the meats and animal-based products (milk, cheese, eggs, leather, wool, etc.) to which we have become accustomed.

Currently, most of the price pressure comes from incentives to produce biofuels and customs duties imposed by the industrialized countries. The United States alone spends $7 billion each year subsidizing the domestic production of bioethanol. These market distortions only aggravate the problem, displacing other crops in favor of corn for biofuel and discouraging more efficient food producers, without any benefit to the market or the environment.[28]

Moreover, the increase in the market for biofuels will gradually lead to a tight correlation of crude oil prices and food prices.[29] Indeed, in a context like this, high prices for crude oil will cause high prices in the agricultural sector (at least the part of it that uses fossil fuels) and create an incentive for the production of biofuels, with a consequent reduction of land harvested for food. Warnings of these kinds of future repercussions have already come up, and they appear even more disturbing considering that the development of biofuels is still in its infancy.

Consider the humble Mexican tortilla, familiar to the patrons of American sports bars and Southwestern cuisine restaurants. It is also the basic food staple of Mexico, vital for those living in poverty, which is more than 50 percent of the population. It is obtained from white corn, which Mexico largely produces domestically. The country imports large amounts of yellow corn from the United States to use as feedstock for animals.

During 2006, the price of a Mexican tortilla almost doubled. Throughout the year, the cost of American yellow corn increased drastically because of the massive use of corn to produce bioethanol, so that Mexican importers were forced to pay $4.20 per bushel compared to $2.80 a few months before.[30] They began to purchase white corn on the domestic market, because it cost less, and to use it for animal feed. This phenomenon put the tortilla makers in competition with the animal breeders for the purchase of white corn, with the result that the price of white corn went sky high. Faced with protests in the streets and a crisis that threatened to grow uncontrollable, Mexican president Felipe Calderón was forced to impose price controls on cereal products.

Similarly, a small boom in the production of biofuels in China during 2006 and 2007 helped raise the price of pork, one of the most widespread foods in that Asian country. Between January and June 2007, pork prices went up more than 40 percent. The driver here was the increase in the cost of animal feed, largely based on cereals that the Chinese had begun to use on a broad scale to produce biofuels. The same phenomenon occurred in the United States, where the increase in the demand for ethanol in 2007 generated price increases for bread (up 10.5 percent), milk (up 19.3 percent), meat (up 5 percent), and derivative products.[31]

The role of biofuels during the 2007–2008 rise in world prices for food and animal feed remains a topic of debate. Estimates average between 15 and 25 percent of the total price increase (from as low as zero to as high as 75 percent).[32] A paper prepared in 2009 by the Federal Reserve Bank suggests that biofuels had a significant impact on individual crop prices (corn, sugar, barley, and soybeans) but a smaller impact on global food prices.[33]

However, the problem of food displacement for the production of biofuels hangs in the background like menacing clouds that could suddenly become a tornado. In fact, with an oil price higher than $70 per barrel, most farmers have an extraordinary incentive to produce biofuels instead of food. Both the Food and Agricultural Organization of the United Nations (FAO) and the OECD have warned about the danger of this cloud. A rapid increase in the biofuel sector could lead to an increase in the price of food between 20 and 50 percent in the coming decade.[34]

The Future

What about the question raised by the OECD: Is the cure worse than the disease? Certainly, the potential contribution of biofuels using current technology has proven to be very limited and hampered by many problems. These will become more obvious as production increases. Considering our present level of knowledge, I see only two roads to a viable future for biofuels.

The first is to identify areas of the planet that naturally lend themselves to large-scale cultivation of ideal biofuel crops without modifying the overall ecosystem. This road has the advantage of offering a real opportunity for development to countries with great difficulties, especially in Africa, as long as the industrialized countries do not resort to devices to protect national crops. In other words, future exports from poorer countries should not be burdened by customs duties and unfair taxes, one of the shameful ways by which the West penalizes the few products that these countries can sell at competitive prices.

The second road looks further into the future for new production frontiers using cutting-edge research. Focused research on nontraditional crops and efficient processes could at least partly resolve the current problems of biofuels in terms of quality, cost, and competition with nutritional uses.

One solution may be near at hand: using woody-cellulosic materials to produce second-generation biofuels. Unfortunately, the technology still has a long way to go to prove itself reliable and profitable on a large scale. Both the biochemical and thermochemical processes have reached the demonstration stage, but there are significant technical barriers yet to be overcome.

There are no shortcuts in overcoming the technical barriers, as a recent scandal has taught us the hard way. Second-generation biofuel received special attention from the U.S. Environmental

Protection Agency, which set a modest production target of 100 million gallons (380 million liters), giving producers a tax credit of $3 per gallon. The modesty of the target shows that the industry is still in its infancy. It is now practically certain that the target will not be met after fraud by one of the production companies was exposed. In the story published in *Scientific American,* University of Massachusetts chemical engineering professor George Huber wrote: "There are no magic processes for conversion of biomass into liquid fuels. . . . If something sounds too good to be true, it probably is not true."[35]

Advanced biofuels do nothing to alleviate the high costs of production we see with first-generation biofuels, especially in the logistics chain. Harvesting, treating, storing, and delivering large volumes of suitable biomass feedstock year-round to a biorefinery has a huge impact on total biofuel cost. In fact, the energy density disadvantage is aggravated by the cost of biomass collection and delivery from distributed regions to central plants. Logistical costs are often not considered carefully enough when second-generation biofuels are studied.

Land requirements remain a clear disadvantage compared not only with conventional oil products but also with those derived from nonconventional oil. In fact, the land area required to produce second-generation biofuels is more than a thousand times greater than the land required to produce the same amount of energy from oil sands or oil shale.[36]

According to the International Energy Agency,[37] advanced biofuels are competitive with gasoline or diesel when the crude oil price is between $100 and $130 per barrel. By 2030, even if production costs were to decrease, advanced bioethanol and biodiesel would still not be competitive if the price of oil were below $70–$80 and $80–$100 per barrel, respectively.

Commercial-scale crops are unlikely to be widespread before 2015 or 2020. Therefore, it remains doubtful that second-generation biofuels will make a significant contribution to meeting the demand for transportation fuel by 2030. Most important is the fact that many of these biomasses are more readily used for heat and electricity (e.g., in home furnaces or in the industrial thermoelectric sector), where they can provide higher energy efficiency and lower greenhouse gas emissions than in complicated and expensive transformation processes for the transportation sector.[38]

Another path for biofuels would be more revolutionary, and thus riskier. It would involve shifting attention radically from the current focus on traditional crops as the sources for biofuels to the feasibility of using microorganisms such as algae, phytoplankton, bacteria, and fungi. These microorganisms feed on the nitrogen and phosphorus found in wastewater and on carbon dioxide, so they could also dispose of substances that damage the environment and affect the climate.

In the laboratory, algae productivity is higher than that of the best vegetable crops, but the laboratory is a long way from the gas pump. So far, the most promising yield achieved from algae has been 0.25 barrels of oil per hectare-day (bbl/ha-day), or about 14,500 liters per hectare-year (L/ha-year). For comparison, Brazilian sugarcane yields 6,000–8,000 L/ha-year.[39] In the long term, it should be possible to achieve a productivity of about 58,000 L/ha-year.

The potential advantages of algae are vitiated by a series of problems that could narrow their real prospects. First, to achieve such high productivity, algae require a supply of enriched CO_2, such as power plant flue gas. The supply, transfer, and use of a high-CO_2 source is a central issue in this technology. Second, current projections are based on laboratory experiments that are modest in number, scope, and achievement. The laboratory studies cannot substitute, predict, or replicate outdoor cultivation experiments. Therefore, the reality is that practical microalgae oil technology does not yet exist, and all aspects of algal oil production still require considerable research and development.

The biological issues are the major challenges. In this field, the main goal is to select the most suitable algae strains in terms of productivity, oil content, and culture stability (e.g., susceptibility to temperature and ability to survive zooplankton invasion). Another important problem is controlling the reproductive processes of the microorganisms themselves. This is almost impossible without confining them to segregated areas, which greatly limits their productivity and therefore their potential. In any case, microalgae oil technology deserves a serious effort in terms of research and interdisciplinary collaboration.

Regardless of the road taken, I find it difficult to imagine that biofuels could make a significant contribution to the enormous demand for energy worldwide. They might lighten the pressure on petroleum in the coming decades by a modest percentage, but they will not free us from it.

Statistics on Biofuels

Table 7.1. Ethanol production (Top 3 countries and world total, 2008)

Country	*Production (millions of tonnes of oil equivalent)*
United States	17.7
Brazil	12.4
Europe	1.5
World total	**33.2**

Table 7.2. Biodiesel production (Top 6 countries and world total, 2008)

Country	*Production (millions of tonnes of oil equivalent)*
Germany	2.8
United States	2.2
France	1.6
United Kingdom	0.8
Brazil	0.7
Italy	0.6
World total	**10.5**

CHAPTER 8

Gone with the Wind?

Around the world, there is a new "wind rush" to generate electricity. It is making wind the fastest-growing energy source among renewables and a prominent focus of energy policies. The two countries where wind energy is escalating most rapidly are the United States and China, albeit starting from a small base.[1] It is still uncertain whether this boom will make a dent in the vast amount of primary energy that will be needed between now and 2030. Today, wind power covers only 0.1 percent of total primary energy consumption, and about 1 percent of electricity production.[2]

Certainly, its theoretical potential is much greater and its costs are even competitive with the costs of fossil fuels on a life-cycle basis. Consider, for example, that wind produces no carbon dioxide, other greenhouse gases, or local pollutants. Unfortunately, the theoretical potential and advantages of wind power run up against a series of problems that could limit its real prospects.

First, wind energy cannot provide amounts of energy comparable to fossil or nuclear fuel. The low power density of wind makes its development disrupt the natural environment, because it requires hundreds of wind turbines (with blade diameters exceeding 260 feet [80 meters]) deployed across many miles to barely match the capacity of a single coal- or gas-fired power plant. The impact of wind turbines on the skyline, along with their noise, the electromagnetic disturbance, and the threat they pose to birds, has caused strong negative reactions from local communities and environmental groups, blocking many projects.

In many countries, windy areas are limited. As with oil, coal, and natural gas, the places where wind resources can be exploited on a significant scale are quite concentrated. In other cases, even where windy areas are available, they are not suitable for electricity production, either because they are too far from consuming markets or because installing wind turbines would unacceptably disfigure the natural setting.

Second, like the sun, the wind is a variable energy source; its availability depends on the whims of nature. This fact has several critical implications. Because storing electricity remains a dream (except for small amounts in batteries), what the wind can produce may be unavailable when it is most needed. Added to the distance from consuming markets, this variability brings uncertainty and calls for costly improvements to wind power systems. Particularly, they require enhanced transmission and distribution systems, along with an alternative energy source to produce electricity when the wind slackens.

This mixed picture has not prevented some European countries from developing wind power for a significant part of their electricity needs. The most relevant case is Denmark, which relies on wind power for about 20 percent of its electricity consumption. The real issue will be to see whether the successful Danish experiment can be replicated globally, or if it will represent a real option for only a few countries.

An Old Story Revisited

Like other renewable energy sources, wind power has a long history. The Persians harnessed the wind for mechanical energy 1,500 years ago. The first wind turbine capable of driving an electric generator was built in 1886 by American Charles F. Brush in Cleveland, Ohio, the city where, a few years earlier, John D. Rockefeller had started what became the modern oil industry. Brush's turbine had a tower 59 feet (18 meters) high and a rotor 56 feet (17 meters) long, consisting of 144 fixed vanes and a hub. It was connected to a generator that powered 100 incandescent light bulbs. It was a marvel of its day and worked incredibly well for more than fifteen years, until Cleveland received the first centralized electric power system.

The American scientist had introduced the world to the idea of generating electricity from the wind, opening a completely new prospect for humanity. Almost at the same time as Brush, another inventor across the world was taking wind energy to even higher levels. His name was Poul La Cour, and he gave his country, Denmark, absolute first place over this source of energy, which earned him the sobriquet of the "Danish Edison."

Thanks to La Cour, by 1906 the Scandinavian country had installed forty wind turbines, the result of his original research. He also had an important impact in Germany, which in the early twentieth century temporarily assumed leadership in the development of wind systems. During the 1920s, technical evolution and the growth of the size of turbines seemed to presage a brilliant future for wind power. They spread rapidly in the United States and northern Europe, particularly in rural areas that were difficult to reach with centralized electrical power systems. Unfortunately, a tornado was about to rip through the wind energy sector.

In 1936, the United States passed the Rural Electrification Act, which subsidized the distribution of electricity to farm cooperatives in isolated rural areas. This led to the triumph of centralized generation systems over distributed systems.

Centralized systems consist of large power plants (at that time, mainly fed by hydropower systems or coal-fired plants), which produce vast amounts of electricity in one place and distribute it to a vast area. *Distributed systems,* by contrast, produce smaller amounts of electricity where it is needed, sometimes for a single consumer or a cluster of consumers. Wind power belonged to this latter category, and so it suffered the consequences of the new law. Research collapsed, and businesses manufacturing wind turbines began to die out, disappearing completely by 1957.

Only Denmark and Germany continued to keep wind energy alive, continuing research and development during and after World War II. By the end of the 1950s, they had realized the first prototypes of modern wind plants.

The most important figure in this effort was another Dane, Johannes Juul. In 1956, building on the trials conducted in Denmark and Germany, Juul developed the first three-vane installation, driving a rotor 79 feet (24 meters) in diameter with stall regulation. *Stall regulation* allows the power of the installation to be maintained at a constant level as the flow of wind over the vanes changes. Once again, however, fate played against the wind.

The low prices for crude oil during the twenty-year period from 1950 to 1970 led to the golden age of oil, allowing it to penetrate into every aspect of modern life, including the production of electricity. Meanwhile, coal became even cheaper and thus more convenient. This left little room for other sources of energy except nuclear power, which benefited both from military research and government financial support.

Already confined to northern Europe, wind energy only reemerged from this eclipse with the oil shocks of the 1970s. The resurgence started in the same European countries that had always believed in wind energy, Denmark and Germany. By now, Denmark had assumed technical leadership in wind energy, but a real boom happened in California, where it was called the "Wind Rush."

There were two main reasons behind this unexpected phenomenon. On the one hand, in the late 1960s and early 1970s, California had been setting the pace for the world in environmental awareness. On the other hand, while the state led the United States economically, its primacy rested on a fragile energy foundation: it was rich in crude oil that was too expensive for electrical power, but it lacked significant deposits of cheaper sources like coal or natural gas to produce electricity.

Wind (and solar) energy benefited from the convergence of these factors, and from generous state incentives and federal tax credits. Danish technicians and entrepreneurs flocked to the West Coast of the United States, providing the know-how for the development of the California wind power sector. The technology was still in its infancy, with turbines having a rotor diameter less than 65 feet (20 meters) and a capacity of only a few dozen kilowatts. Because of these limitations, entrepreneurs installed them in large arrays, which soon became known as "wind farms." By 1985, California had installed 1.2 gigawatts (GW) of wind power, then corresponding to 90 percent of world's total.

The boom did not last long. In 1985, subsidies for alternative energy sources were discontinued, dealing a serious blow to the new industry. Then the next year, the collapse of oil prices drew the curtain on the California experiment.

When wind energy finally stirred years later, it changed direction again. In the mid-1990s, northern Europe again undertook wind development, making important technological advances and building wind farms not only on land but also offshore, where the lack of obstacles assured a better energy yield. The boom has not stopped, spreading outward from that small band of countries that had kept alive the hopes of clean and economical wind energy for more than a century.

How Is the Wind Exploited Today?

According to the Global Wind Energy Council, at the end of 2008, wind turbines throughout the world had an installed electrical generation capacity of 121 GW, compared to generation capacity from all sources of more than 4,500 GW. In other words, they represented more than 2 percent of the global electric capacity. The amount of electricity actually generated by the wind was lower, a very modest 1 percent of that produced worldwide (230 TWh from wind power, compared to 20,300 TWh from all sources).[3]

Data about installed power can be deceiving. If the same 121 GW of installed power had been from gas- or coal-fired plants instead of wind turbines, the electrical energy produced would have been three times greater. The variable nature of the wind makes it impossible to exploit fully the installed generating capacity of the wind turbines. It is like owning an automobile but only having gasoline at certain times of the year.

On average globally, wind turbines work at full load for the equivalent of 1,500–3,000 hours of the 8,760 hours in a year. Therefore, they have a *capacity factor* between 20 and 40 percent, with some exceptions.[4] The broad variability depends on the windy conditions in a specific area. By comparison, the capacity factor for nuclear power is greater than 90 percent, which means that it is working almost continuously. The capacity factor for gas- and coal-fired plants is more than 80 percent.

Looking at it this way, the numbers for wind power seem so insignificant that they raise doubts about its real ability to provide hope for the energy future of humanity. However, we must consider other aspects.

In just a little more than ten years, the installed wind power capacity of the world has increased twenty times, an impressive growth rate. In 2007 and 2008 alone, it increased by about 50 GW, surprising all experts and observers. Considering this rate of growth and the projects under way in many parts of the world, in a few years these numbers will be probably be outdated. It is true that when starting from such a modest basis, percentage increases of 30 percent or even 50 percent change the overall contribution of wind energy to the world energy mix by very little. However, it would be a mistake to underestimate the growth potential of the wind.

Europe still represents the only area where wind power has assumed a significant role in the production of electricity. Denmark is the main beneficiary of wind-generated electricity (20 percent of the country's consumption). However, in 2007, wind also made a noticeable contribution in Spain (10 percent), Germany (6 percent), and Portugal (8 percent).[5] Until 2008, Germany led the world in terms of installed wind power capacity.

The wind has gradually lost its European focus. In early 2009, Europe owned just half of the world's installed capacity, down from around 75 percent in the early 2000s.[6]

In 2008, the United States overtook Germany for first place in installed wind power capacity, with Texas playing a leading role.[7] In Asia, the growth of wind power has also been very strong, and China could shortly threaten the new U.S. leadership.[8]

Worldwide, wind power systems are still highly concentrated geographically, much more so than reserves of oil and natural gas. More than 70 percent of the wind-generating capacity is concentrated in only five countries: the United States (25 GW), Germany (24 GW), Spain (17 GW), China (12 GW), and India (10 GW).[9]

Because wind power emits no greenhouse gases or pollutants, its development has been supported by public subsidies and other incentives, along with provisions that require traditional utilities (mainly power companies) to produce part of their electricity from renewable sources. Formulas of this sort are spreading. For example, in September 2007 the Chinese government issued a law that requires large power companies to generate at least 3 percent of their electricity from renewable sources (excluding hydroelectric power) by 2010. The share will gradually increase to 8 percent in 2020.

Besides political support, significant technical advancements have driven wind power development, although the basic structure of wind turbines has not changed much. The vast majority of them are of the three-vane type first tested by Juul during the 1950s. The rotor transmits the wind energy through a system of shafts connected to an electrical generator.[10]

For the same wind speed and generator power, electricity produced by a wind power installation depends on the volume of air captured by the vanes, and thus on the length and the design of the vanes themselves. The most progress in recent years has been made on these two elements.

In the mid-1990s, the largest wind-powered generators had a capacity of about 500–600 kilowatts, with a rotor diameter of about 165 feet (50 meters). Today the rotors on the biggest wind turbines can be 360–400 feet (110–120 meters) in diameter, and they are installed on towers more than 330 feet (100 meters) tall. The generators connected to these turbines on average are larger than 1 megawatt (MW) and the biggest ones can be up to 6 MW. Of course, the size of the turbines and the power of the generators that can be installed depend on the availability and the speed of the wind in a specific area, a subject that requires a brief digression.

To make it easier to understand, it is enough to remember that wind-powered generators need a wind speed of 4–5 meters per second (m/s) or about 8–9 knots. This is called the *startup speed*, and it is the minimum speed to overcome internal friction and inertia. There are also *overspeed controls*, safety devices that stop the generators at speeds above 20–25 m/s (40–50 knots) to avoid excessive vibration to the structure. In other words, they cannot be used in seriously bad weather such as nor'easters, gales, and hurricanes.

The power of the wind is directly connected to its speed and can be calculated as the cube of the speed with a correction coefficient (because the proportional increase in power is greater at higher speeds).[11] By way of example, in a windy area, one can count on an average wind speed of about 6 m/s (12 knots), which corresponds to an average wind power of about 200–300 watts per square meter, which must be intercepted by the vanes of the rotor. The yield of the rotor in converting the kinetic energy of the wind into mechanical energy is about 45–50 percent. The generator then transforms this into electrical energy, with a final yield of less than 40 percent. In places like the North Sea, the average wind speed is twice that and can drive generators with greater power output.

Because of friction close to the Earth's surface, the wind speed increases farther from the ground. For this reason, wind towers are becoming ever taller, and ideas resembling science fiction are being considered for building wind power systems suspended in the sky (e.g., kite wind generators).

The drag on the wind also varies depending on the different surfaces over which it flows. This drag is classified in a *roughness* scale for surfaces, from the lowest (class 0 = open sea) to the highest, or most wind resistance, such as forests and villages (class 3) or cities with skyscrapers (class 4). This explains why offshore wind farms are being designed all over the world. In Europe, offshore wind already accounts for nearly 2 percent of total installed wind capacity.[12] In 2008, Great Britain overtook Denmark as the world leader, with around 600 MW. Britain plans further expansion, which has led to more than one controversy.[13]As mentioned in the chapter on natural gas, Great Britain risks running short on power in the next few years also because of over-enthusiastic support for wind power development, which is appearing very difficult to carry out.

The prospect of offshore wind development is encouraging other important technological advances. One of the first goals is to increase the power capacity of wind turbines to 10 GW, using different systems and materials to withstand the strength and variability of winds. For example, scientists and engineers are studying new versions of a two-blade turbine on a vertical axis, which could be much cheaper than the equivalent three-blade turbine. Another target is gearbox durability, using systems than may even eliminate the need for gearboxes.

A third element that could support significant development of wind in the future is its enormous theoretical potential for generating electricity, up to several thousand terawatt-hours per year. This is comparable to the entire amount of electricity consumed worldwide. In July 2008, a report by the U.S. Department of Energy (DOE) calculated that the United States has a theoretical potential of around 600 GW of wind power that could be exploited at a cost of around 6–10 cents per kilowatt-hour.[14]

These numbers lead to the fourth reason for the potential success of wind power: cost. Modern, medium- to large-size onshore generators cost on average $1,800 per kilowatt installed, with a very low operating cost (2–3 percent of initial investment annually).[15] Because wind turbines operate irregularly, the cost of a wind-powered kilowatt-hour strongly depends on the specific capacity factor of the installation.

Under the best conditions, wind power is already quite competitive with fossil fuels, generating electricity at a cost of about 6 cents per kilowatt-hour. In areas that are not exceptionally windy (an average wind of 6–7 m/s), the cost of wind-powered electricity is still between 50 and 100 percent higher than fossil fuel. In those cases, wind power could not compete without public subsidies or other support. It is interesting to note that during California's Wind Rush, a kilowatt-hour from wind (under the best conditions) cost about 30 cents.

For all these reasons, several countries in the world have ambitious plans for wind power development. In its July 2008 report, the DOE has explored the possibility of achieving 20 percent of the electricity supply from wind by 2030. Even if substantial challenges and impacts are associated with this scenario, the DOE found that the huge American wind resources would be adequate to achieve the target at a reasonable cost.[16] In President Barack Obama's new green agenda, wind power has a major role, and a new policy of subsidies and incentives has been established to support its development.[17]

In 2008, the Chinese National Energy Administration highlighted wind energy as a priority for diversifying China's energy mix. In Europe, new regulations and incentives were issued to support wind power investment, and a program was launched aimed at developing more than 100 GW of installed capacity by 2020.[18]

With strong political support, wind seems to have entered a golden age. So what is there to stop it?

Blowing against the Wind

To start with, the major problem with wind power is its erratic nature. This brings about several complications and hidden costs. As we have seen, the wind blows at full speed about a third of the time (more or less) in the windy regions of the world. This leaves wind power generators much less productive than their rated capacity. The amount of electricity that can be obtained with wind-powered installations is relatively modest and is often available when it is less useful.

This intermittence causes other problems. First, the electrical grid must be capable of withstanding significant power swings, which do not occur with on-demand energy sources. This problem becomes critical when installed wind capacity exceeds roughly 10 percent of total electricity production, because on average for one-third of the time, there will be too much wind-powered electricity, and for two-thirds of the time, there will not be enough.

To face this issue, transmission lines and distribution grids must be adapted or even rebuilt as the share from wind power increases.

It would be especially preferable to base them on high-voltage direct current (DC), instead of the alternating current (AC) that we use today. Their architecture also has to become more complex and flexible to deal with the erratic inputs of energy. In turn, grid operators would need to be able to activate other energy sources promptly when the wind slows.

This complication would be even greater in a system based on distributed energy, where many self-producers (for example, a single owner of a wind turbine or of a solar photovoltaic system) could sell or buy electricity to the grid as available or needed. These kinds of sophisticated grids involve the extensive use of a variety of digital technologies for monitoring and controlling the flows of energy in systems where the contribution of renewables is significant. Because of their complexity, they have come to be known as *smart grids*. Unfortunately, most analysts usually underestimate the very high overall cost of having high-voltage DC transmission systems integrated with smart grids. High-voltage DC smart grids will probably provoke a backlash from many consumers. Smart grids require a broad consensus among the different operators who will use them, and their vulnerability to hacker attacks poses a security problem.

The second problem with wind intermittence is that it requires a backup energy source. The most suitable on-demand source is natural gas, because gas-fired power plants can be switched on and off quickly, unlike nuclear and coal plants. This involves another often unrecognized cost.

Consider the simplified case of the owner of a gas-fired plant serving as swing producer, that is, the gas-fired plant switches off when the wind blows. If the gas plant is used only for a limited amount of time, it may not pay for itself. The gas-fired plant operator also needs a flexible supply of natural gas (e.g., access to stored gas), which has a cost as well. Who pays for these services, which are essential to the reliability of overall electricity supply? Naturally, it is the consumer, but most calculations about wind power cost do not include this critical aspect.

Another significant cost of wind power comes from the need to connect remote windy regions to the major consumption areas. This is the case in most of the world. For example, in the continental United States, the bulk of wind resources are located in the Great Plains and Midwest, while the big consuming centers are spread along the two coasts. China's wind resources are in the remote north and northwest (e.g., the Gobi Desert), while most electricity is consumed in the southwest. High costs and long lead times for this kind of transmission infrastructure could pose a significant obstacle to the timely development of new wind capacity.

In the United States, power network development and management could turn out to be a major barrier, especially because the country is divided into more than 140 balancing areas with a very low level of interconnection. It has been estimated that if the United States wants to get 20 percent of its electricity from wind, it will need to build a new transmission overlay with 15,000 miles (24,000 kilometers) of new extremely high-voltage lines, at an estimated cost of $100 billion.[19]

Calculations about wind power potential often do not take into account the fact that many windy areas feature natural environments that would be damaged by the construction of wind farms. This brings us to the final problem concerning wind power: public acceptance.

In a way, this issue derives from the modest power density of wind power. Low power density makes it necessary to build arrays of wind turbines for miles and miles to obtain quantities of electricity that are comparable to those produced by small coal-fired or gas-fired power plants. A practical example of this principle is much more catching.

Building a wind farm with the power of one average-size thermoelectric plant (500 MW) would require deploying 250 two-megawatt wind generators (the most common size) across about 100 miles (160 kilometers). Even under the best conditions, taking advantage of a windy offshore area with larger wind generators (5 MW) would require deploying 100 of them over 17 square miles (45 square kilometers). By comparison, one ordinary power plant fired by coal or natural gas on land uses less than 0.4 square miles (1 square kilometer).[20]

This low power density makes wind farms environmentally and visually disruptive. Wherever they appear, they greatly modify the natural setting, usually for the worse. This may be tolerated for the environmental advantages of wind power, but the visual aspect has proven to be a powerful obstacle to the spread of wind energy. Along with it come the problems caused by working of the turbine blades, such as electromagnetic interference, noise pollution, and the mortal danger they pose for certain bird species. Over time, these problems have found partial solutions, but they remain among the principal motives for opposition to wind farms by many environmental groups, who have already succeeded in stopping several wind farm projects around the world.

We may draw a mixed picture of wind power from all these complications. Its effective potential is greatly limited by natural, hidden costs, and social implications. Its development requires a substantial change of existing systems for the transmission and distribution of electricity. So, what is its real future?

The Future

Navigating the pros and cons of wind power and understanding its future is not easy. For wind and renewable energy supporters, this exercise will be relatively important. Skeptics believe that sooner or later the Wind Rush will die out, leading many investors to bankruptcy.

In my view, wind power will continue to be the fastest-growing source of energy in the future among renewables, and it will make a contribution to the primary energy basket of humanity by 2020 or 2030. However, the extent of this contribution is uncertain.

How far wind power will extend will depend on the steady, long-term ability and willingness of governments to subsidize wind energy, to embark on massive plans to build new transmission networks, and to devise rules that allow for the remuneration of those operators who will provide reserve capacity. At the same time, the diffusion of wind farms will depend critically also on public acceptance, which we cannot take for granted.

Given this context, my feeling is that the success of wind power will also limit its real potential. In other words, I fear that most of the problems of this green energy will emerge in parallel with its development on a larger scale, and this will dampen its rate of growth. Furthermore, public funding and subsidies are not infinite; after the excitement cools and the first problems arise, funding generally tends to be curtailed.

Consequently, I believe that wind energy will play a significant role, although a niche role, in satisfying the energy needs of our planet. It will be an advantageous solution in some localities, but it will be unable to challenge the large requirement that characterizes our near-term global demand for primary energy.

Its contribution to global demand could amount to perhaps 2–3 percent by 2030, and to a more substantial percentage of electrical production. I doubt that the Danish achievement could be replicated on a global scale.

The peculiar characteristics of Denmark in terms of wind resources resemble those of Brazil in terms of biofuels production. Simply put, these two countries are unique, as Saudi Arabia is unique in terms of oil endowment. Furthermore, Denmark is interconnected with its neighbors for its energy flows. It can sell wind power overproduction to Norway (for easy storage in pumped-storage hydroelectric plants) and can take back energy from interconnected markets when the wind farms do not work.

Unfortunately, the world is not Denmark, and the future of wind power must confront this naked reality.

Statistics on Wind Power

Table 8.1. Installed wind power electric generation capacity (Top 10 countries and world total, 2008)

Country	*Installed capacity (GW)*
United States	25.2
Germany	23.9
Spain	16.8
China	12.2
India	9.6
Italy	3.7
France	3.4
United Kingdom	3.2
Denmark	3.2
Portugal	2.9
World total	**120.8**

Table 8.2. Wind power electricity production (Top 10 countries and world total, 2008)

Country	*Production (TWh)*
United States	52.0
Germany	40.4
Spain	31.5
China	18.5
India	12.5
United Kingdom	7.1
Denmark	6.9
Portugal	5.7
France	5.7
Italy	4.9
World total	**216.2**

CHAPTER 9

The Sun Also Rises

In fairness, I must say first that I consider solar energy to be the great energy hope for the future of humanity. Each year, the Earth receives from the sun thousands of times more energy than humanity consumes worldwide. In theory, capturing just a minimal part would be enough to supply all the electricity we need, without changing the climate, damaging the environment, or exhausting the resource.

Yet today, solar energy does not even cover one one-thousandth of the primary energy consumption of our planet. Even in those countries where it generates the most electricity (Germany, Japan, and the United States), solar numbers are insignificant, less than 1 percent of all primary energy. Even these meager percentages would not be possible without public subsidies and ad hoc regulations that the three countries have put in place to encourage the use of solar energy.

So far, the sun has escaped every stratagem for effective capture, although it has the highest power potential of all renewable sources. Vast areas are required to obtain relatively modest amounts of energy, and the technology available today is still too expensive for large-scale exploitation. Nevertheless, its greater power potential compared to its clean competitors and the encouraging prospects for improving the technology give hope for its future exploitation. This hope may not materialize in two or three decades, but further out, the sun could offer us a realistic alternative to fossil fuels.

Another Old Story Revisited

The first modern attempts to harness solar power go back to the nineteenth century when the Industrial Revolution was in full swing. At that time, the first systems to produce electrical energy from high-temperature steam resembled the concentrated solar thermodynamic installations of today.[1]

The most interesting result of the experiments of the day was a machine that Frenchman Augustin B. Mouchot presented at the 1878 Paris World Fair more as a curiosity than as a useful device. About 1910, German engineer Frank Schumann built an installation in Egypt, consisting of long, parabolic concentrators that focused solar rays on tubes of darkened glass, connected to a heat accumulator (a large tank covered with insulating material that collected the hot water). The rapid rise of petroleum and the death of Schumann himself during World War I caused his research and installations to be abandoned.

The possibility of transforming light into electricity by interactions with certain materials (the *photovoltaic effect)* was discovered almost accidentally. As far back as 1837, French physicist Edmond Becquerel noticed that some electrolytic solutions developed weak electrical voltages when illuminated. Although the phenomenon was not completely understood or explained, in late 1870 the first practical applications exploited it using selenium solar cells, developed by English scientists William Adams and Richard Day.[2]

The photovoltaic mystery would not be unraveled until the early twentieth century, by the research of Max Planck and especially Albert Einstein, who won his Nobel Prize (1905) for explaining the photoelectric effect (and not for the theory of relativity, as many people think). Their research did not have practical effects immediately.

Photovoltaic devices with significant conversion efficiency arrived in the 1930s, with the production of new versions of selenium cells.[3] The modern era of solar power took a major step forward in 1954 when American researchers at Bell Laboratories (where the transistor had been invented) developed the first silicon solar cell with an efficiency of 6 percent.

Developed on a small production scale, the first Bell cells faced enormous marketing difficulties because of their prohibitive cost. Until the 1970s, application was limited to electrical power supplies for satellites and spaceships, for which there were no valid alternatives. This was their lucky break.

During the space race between the Americans and the Soviets, the U.S. government heavily financed photovoltaic research programs, giving rise to specialized industrial initiatives. Solar technology improved, and the cost of the cells dropped noticeably, although it remained impractical outside of space and military applications. The first Bell solar cells with 1 watt of power in 1956 cost about $380; those used by *Skylab 1* in 1973 cost about $100, which was about two hundred times the cost of conventional electricity at the time. On this basis, there was virtually no room for widespread industrial application. Without the revenues from the space programs of Washington and Moscow, modern solar energy would never have taken off. Instead, the space race led to applications on the ground.

Paradoxically, an important contribution came from a petroleum company, Exxon. In 1969, convinced by an internal task force that energy prices (of oil in particular) would be much higher by 2000, the queen of the "Seven Sisters" established a company (Solar Power Corporation) for the purpose of developing solar cells for terrestrial use. By 1980, these initiatives by oil companies and others lowered the cost of solar cells to about $20 per watt, still about forty times that of conventional electricity. The first solar cells for terrestrial use were produced essentially for remote applications, in areas that could not be reached by the electrical grid. This created a niche market, sustained largely by the oil companies for the electrification of offshore oil platforms, anticorrosion systems for oil wells and oil pipelines, communications networks, and rural villages.[4]

This first phase of modern development allowed solar power to take advantage of the first oil shock (1973), which seemed to change its prospects profoundly. The very high prices for crude oil, the general fear of continuing blackmail of the consumer countries by the producer countries, and the widespread fear that the world would soon run out of black gold created real interest in all alternative energies. In short, a flood of government research programs, public subsidies, and financing sprang up, as well as private initiatives, mainly in Europe, the United States, and Japan. After the financing phase tied to the space program, this became the second major phase in the development of photovoltaic power, which expanded its application to civilian and industrial uses along with additional improvements in technology, efficiency, and cost.

The collapse of oil prices in 1986 and the end of the fear of the 1970s caused drastic reductions in solar energy initiatives. Nevertheless, a small circle of countries, primarily Germany, Japan, and the United States, have continued to support research, allowing solar power to continue evolving, albeit more slowly.

What Is Solar Power and How Is It Exploited Today?

Sunlight, the radiation that each day allows our planet to live, is a direct source of energy. Unfortunately, it cannot provide electricity directly. To do this, it must be captured and transformed (the word "transformed" is not exactly right, but I use it to facilitate understanding). Although the energy from radiated light can be used directly to heat a fluid (e.g., water), an intermediate device is needed to obtain electricity. The device can work in one of two ways. It can transform the light energy into thermal energy, forming steam to drive a turbogenerator, or it can transform the solar light directly into electricity.

Today, there are at least three major systems for capturing solar radiation and transforming it into thermal or electrical energy. The first of these is the simplest: low-temperature *thermal solar power*. It provides hot water and hot air through elementary devices: glass, metal or plastic collectors (panels), and a reservoir tank for the hot water. Depending on their complexity, they can cost between $1,500 and $6,000 for a complete installation capable of providing hot water for a medium-size home and a family of four people.

More complex and expensive versions of such systems can heat and cool large buildings. It would be good to see this technology spread, but it still needs to mature commercially. Thermal solar power stops here, because it does not produce electricity. Therefore, its advantages and its potential are limited.

The high-temperature *concentrated solar power* system is more complex. This technique for exploiting solar energy is based on *collectors* (glass or metal mirrors), which capture solar radiation and concentrate it toward a *receiver*. This is usually a tube attached to the collector itself or a tower in the center of many collectors. Conceptually, the system is no more sophisticated than a device attributed to Archimedes. However, modern receivers contain special fluids (even air, in some experiments). The fluids heat up to temperatures between 300°C and about 600°C, generating enough steam to drive turbines for electrical generation.

The many applications of concentrated solar power have given encouraging results, but they do suffer from a series of limitations. First, these systems provide significant yields only in areas that are heavily exposed to *direct radiation* (solar radiation that produces a shadow, unlike *diffuse radiation*). The ground needs to be flat, and the systems work better in the desert or in tropical latitudes. Second, they need enormous areas to produce significant amounts of energy.

In order for the system to work best, the collectors must follow the sun, like sunflowers. This brings up the likelihood of mechanical problems and a constant need for maintenance. They also become dirty relatively easily, so they need to be cleaned regularly to maintain efficiency.

These are only some of the problems that raise the cost of electricity from concentrated solar radiation. Different estimates indicate between 17 and 25 cents per kilowatt-hour under the best possible conditions of sunlight (e.g., the California desert). This is about five to eight times as much as it costs to produce the same electricity in a coal-fired plant in the United States, without factoring in the cost of the latter's carbon emissions. Nevertheless, the technology is making progress, and new installations are being designed, built, and completed with very encouraging prospects.

An 11-MW solar tower was recently completed in Seville, Spain,[5] the first new installation in about fifteen years. The 40-MW SEGS VIII and IX in the Mojave Desert of California were completed in the early 1990s. The largest new installation, completed in 2008, is Solar One in the Nevada desert, with an installed capacity of 64 MW covering half a square mile (1.4 square kilometers). This compares to a medium-size conventional power plant of 500 MW, which would cover about 0.4 square miles (1 square kilometer).

This new impetus to the concentrated solar power sector was triggered by the expectation of lower costs and technological progress on scale factors and innovation. Some synergy could be gained from integrating solar-powered plants with conventional electrical power plants, in particular, combined-cycle gas plants, using part of the heat produced to desalinate seawater for agricultural and other uses.

The third system able to capture and transform solar energy is based on the photovoltaic effect. Solar energy comes to us as an array of radiations (popularly called "rays"), each one having a different wave frequency and color. The sum of these constitutes the solar spectrum, which we can see refracted in a rainbow. Devices that take advantage of this phenomenon require special modules composed of many *solar cells*. The critical material for these is super-pure silicon cut into very thin wafers of about 200 micrometers (μm).

Although it comes from the one of the most widespread minerals on Earth (silica is found everywhere in sand and rocks), silicon is the result of a costly and sophisticated refining process that yields what looks like a silvery metal. It has been used mainly in the semiconductor and electronics industries to produce countless components for computers, cellular telephones, flash memories, and so on. Since 2006, silicon for photovoltaic devices has exceeded the demand from all other manufacturing sectors. Combined with the high refining costs, this exceptional demand caused a steep increase in the price of silicon on international markets until 2008.

The silicon photovoltaic solar cell is the most widespread system for producing electricity from solar power today. Unfortunately, it has an insurmountable cost/yield problem that creates an efficiency limitation.

A top-quality commercial photovoltaic module can transform only about 13–15 percent of the solar energy it absorbs into electrical energy, mainly because the materials in the cells capture and convert only a narrow band of solar frequencies. To make this notion simpler, think of the two frequency bands we must protect ourselves from when out in the sun: infrared (IR) and ultraviolet (UV) rays. In a silicon cell, red light (IR) has so little energy that it cannot stimulate the electrons into motion. Although blue light (UV) has more energy, it interacts at internal atomic levels and becomes lost as heat. Unable to convert IR and UV rays into electricity, current photovoltaic systems only work on the remaining energy in the light spectrum.

These and other technical causes limit the practical efficiency of silicon solar cells, so that the cost of a kilowatt-hour of electricity produced by a residential photovoltaic system hovers between four and ten times the cost of a kilowatt-hour from fossil fuels (excluding the cost of dealing with the carbon emissions).

As with concentrated solar power, the wide variability in cost is a function of the solar exposure of the place where the solar panels are installed. In the American Southwest, electricity from photovoltaic cells can cost four or five times that of conventional electricity.

In the central parts of the country, the cost can rise to six or seven times; farther north it can cost ten times more. For this reason, present technology using silicon and other inorganic materials does not appear destined to be the future of photovoltaic solar power.

Let me issue one caveat, however. There are already solar cells with yields greater than 40 percent, but these are laboratory samples and panels used in orbiting space satellites. These conditions cannot be replicated on the ground (direct exposure to the sun in the absence of a filtering atmosphere). The technologies and devices needed make their costs astronomical.

Many laboratories around the world have launched initiatives for the terrestrial use of these cells based on concentrating sunlight on small areas of cells using suitable lenses. This would minimize the cost of these precious materials and leave the collection work to less costly plastic, glass, or metal structures (lenses, frames, etc.). This concept is not new, but it has not yielded much in the way of results—yet.

Prospects are good for this line of research, because it is trying to do better than the traditional flat solar panel. Concentrating structures such as prisms can produce the same amount of energy on a smaller surface. Not by chance, research on these types of systems is financed by the military, mainly in the United States. Miniaturized systems for electrical generation would suddenly resolve a load problem, greatly increasing the operational mobility of soldiers and military vehicles in a war zone.

However, engineering these methods is anything but simple. Just think of the need for perfect alignment between the cells and the lenses or prisms, which must be maintained through the heating and cooling cycles of day and night, or the need to follow the sun with great precision. Concentrating systems of this type have been studied since the early days of photovoltaic technology, but they are only now beginning to show promise.

Notwithstanding its problems, solar technologies (especially photovoltaics and low-temperature thermal solar) are booming worldwide, thanks to government-supported programs. Markets are growing at double-digit rates and have been for more than a decade, largely linked to subsidies in Germany, Japan, some U.S. states, and more recently, other European countries.

As with other renewables, the drivers for these subsidies are environmental concerns and security of supply. A major shift in the types of photovoltaic applications has occurred. They are now terrestrial, connected to existing power grids, and used as supplemental power sources, in other words, expensive, environmentally friendly systems in regions that already have electrical power. Remote and special technical applications obviously still exist, but they are growing at a much slower pace.

In the coming years, it should be relatively easy to improve the efficiency of traditional photovoltaic cells incrementally, as well as to lower their cost significantly. For example, in the United States, commercial photovoltaic panels are already being marketed that achieve efficiencies close to or greater than 20 percent, thanks to mechanisms that partially follow the trajectory of the sun, although they cost much more than traditional fixed panels. We still do not have sufficient data about the real advantages of these panels or about their durability over time. A traditional photovoltaic panel has a useful life of between twenty and thirty years, although its yield drops by a small fraction each year with the normal wear from the weather. Clearly, the durability of the panels and the consistency of their output are critical economic factors.

Most of the effort to get new products on the market is, however, in the *thin-film* technologies, which use very thin layers of active materials, deposited directly on the glass or on a plastic sheet. These technologies use much less material than traditional silicon panels.

Historically, photovoltaic cell costs have dropped almost 20 percent each time world production doubled. This is still happening in spite of the recent silicon supply shortage. The production of polysilicon rose sharply in 2008, providing more than enough raw material to industry. Furthermore, overproduction of solar cells globally and the threat of subsidy reductions and market shrinkage during the economic crisis have helped lower prices since their 2008 peak.

New thin-film products, claimed to cost less to produce, have increased competition. In the last few years, in fact, the product that has gained the most attention is the cadmium telluride thin-film panel, rocketing in production volumes and achieving much faster cost reductions than any other photovoltaic technology.

Other products growing in the market are thin films called CIS (copper indium selenide) and thin films based on amorphous silicon (combined with microcrystalline silicon to enhance performance and stability, the *micromorph* cell). In the first case, operators are hoping to reproduce the very high efficiencies achieved in laboratories on an industrial scale, up to 19 percent. In the latter case, toolmakers operating in other sectors such as the flat panel display industry, are starting to sell turnkey equipment and penetrating the solar panel market.

In 2008, worldwide production capacity of photovoltaic panels was about 7 GW, and market demand was about 6 GW, up more than 100 percent from the previous year.[6] Despite an expected dip in 2009 because of the global financial crisis and other internal factors, the photovoltaic market should continue to grow, spurred by subsidies and incentives to invest in renewable sources of energy.

The situation will be somewhat different for concentrating solar power, still in a project-financing phase but with an aggressive schedule supported by a well-motivated community. The market for low-temperature thermal power will also differ; it has already spread widely without subsidies, for example, in the Middle East and southern Europe. China is the major industrial producer of hot water panels, for its domestic market.

The prospects for solar power are thus encouraging, even though its basic problems are even more restrictive than those of wind power.

The Unsolved Problems

Despite the media emphasis on solar power and the present market boom, the numbers for solar energy should make even the most avid supporter feel uneasy. If the sun offers us so much hope, why does it not cover even a thousandth of our primary energy consumption?

The problem is that we still do not have a cheap technology to capture and transform a significant portion of the enormous amount of energy coming from our star. To this we must add the same problem that hampers wind energy: the erratic availability of the source. Not only is the sun unavailable at night and during overcast weather, but it is also a rarefied commodity in vast areas of the planet. Intermittency raises the same problems in connecting to existing power grids that we have with wind power. The much lower installed capacity of solar power mitigates the impact, and the predictable daily cycle does make the sun more predictable than the wind, but a problem of power surge and fall remains.

To produce significant amounts of usable energy, solar concentration and photovoltaic installations need enormous areas. Compared to the typical 500-MW gas-fired electrical power plant sitting on less than half a square mile, the same capacity from a photovoltaic installation would require between 2.5 and 4 square miles (6–10 square kilometers) of land.

The need for vast spaces may be an objective limit, but we should remember that solar energy requires much less area than other renewable sources, at least under optimal conditions of sunlight. Although its power density is only a small fraction of that of fossil fuels, it is much higher than the power density of wind energy or biomass.

In theory, a surface area comparable to the city of Los Angeles (around 500 square miles [1,300 square kilometers]) could generate *all* the electricity consumed in California in 2008 (264 TWh) using photovoltaic systems. By contrast, if we planted the same area in corn to obtain bioethanol, it would provide only 4 percent of energy equivalent of the gasoline consumed by California drivers in the same year.

About 6,500 square miles (17,000 square kilometers) of photovoltaic systems could replace the entire U.S. annual production of electricity. This is only 0.2 percent of the entire area of the country. The problem is that this operation would be unsupportable from an economic point of view, in terms of both investment and, in turn, the generation cost per kilowatt-hour. Using the present unit cost for end customers ($5 per installed watt for roof-mounted systems), the total cost would be about $11 trillion, not far from the entire 2008 gross domestic product of the United States.

It should also be remembered that an extensive solar panel system would inevitably produce concentrated heat, requiring a large open space around the installation. With bright sunlight, photovoltaic cells can easily reach 160°F (70°C), creating a band of heated air.[7] Locating large concentrations of solar panels near urban centers involves many problems that have not yet been studied in depth.

On a smaller scale, the installation of 1 kilowatt of solar power requires on average 85 square feet (8 square meters) of solar panels, depending on the amount of sunlight. This ratio would mean that an average family, which typically uses 3 kilowatts of power, would need at least 250 square feet (24 square meters) of panels.

The goal of the solar community is to reach a point where subsidies will not be needed. In particular, there is great hope that photovoltaic technology being developed by industrial companies will soon be able to meet *grid parity* (the generating cost at which the technology matches or beats the average consumer cost of electricity). This would make it a true energy supply alternative and spur even faster market penetration.

We must be cautious about what grid parity really means, though, and why it should be considered only an intermediate target. We must consider the peculiar characteristics of solar energy in a specific site for marketing purposes and for accounting. For instance, thin-film producers claim that grid parity may have already been reached for at least one area, the Nevada desert. However, the rest of the United States (and the world) is not the Nevada desert, with its vast endowment of solar radiation. Furthermore, even if the generating cost of photovoltaic electricity matches what consumers would pay their utility companies, the initial capital cost is still very high, so getting a return of the investment will take a long time.

Achieving grid parity is understandably an important target for the solar industry, but it will not trigger widespread use of solar energy. Costs need to be reduced much more substantially, and technologies need to be prepared to meet the "terawatt challenge," that is, to be able to be produce much greater quantities than today.

Large-scale manufacturing is another unsolved problem. This involves both material availability and industrial productivity. The most successful product on the market, the silicon-based solar module, requires a long and complex manufacturing cycle. This may prevent the technology from ever meeting a substantial share of the world's energy demand, even though silicon is an abundant element.

Even the much-advertised and most successful competitor of silicon today, cadmium telluride thin-film technology, will eventually be limited by the availability of rare elements like tellurium, not to mention the toxicity of cadmium components. This technology will not be a long-term candidate for mass production.

Finally, the intermittent availability of solar energy requires a supplemental energy source for nighttime and overcast weather. Sunlight only supplies energy about 25 percent of the time, while fossil fuels can work 80–90 percent of the time. This is not a small problem, and it has prevented the spread of photovoltaic systems to where they might be most needed, that is, to areas off the electric grid.

Currently, isolated photovoltaic systems are used with battery packs similar to those in automobiles or with backup diesel generators. Both solutions are unsatisfactory, because they increase the cost of the installation and require maintenance. This is why most development of photovoltaic technology is taking place where there is already an electric grid.

Photovoltaic power presents an undeniable advantage for industrialized countries. It produces most of its electricity precisely at the time of day when demand peaks, especially during air-conditioning season. Photovoltaic systems can smooth out the peaks and avoid blackouts.

Unlike photovoltaic systems, solar concentration systems can store some of the heat generated in fluid-filled heat retention tanks to continue operating turbines without sunlight. Nevertheless, storing large amounts of energy from an intermittent source is still an unresolved issue. Science has so far achieved only small storage devices such as batteries and accumulators, and a revolution in this field is not yet in sight.

A Feasible Future

Cost, power density, and energy storage represent the critical challenges on which the future of solar power depends. These challenges present two different, but not necessarily exclusive, paths for solar energy.

The first path is the one I call "evolutionary mass production," and it requires achieving two basic goals: low costs and large-scale manufacturing. Put simply, this means setting aside technically complex solutions to achieve higher manufacturing productivity and lower costs. Places for improvement include loosening very tight specifications for optics, alignment, and temperatures (such as those found in some concentrating power systems) and very sophisticated ultrathin silicon photovoltaic cells. This path has worked in the past, for example, when the space-agency specifications for photovoltaic panels were relaxed for terrestrial use and a cost reduction curve started developing.

The second path is the "revolutionary path," which implies one or more major breakthroughs that can change the rules of the solar game. These breakthroughs must occur first in the materials used to capture solar radiation and transform it into electrical power.

This means, for instance, photovoltaic materials capable of absorbing and converting a broader range of solar frequencies (colors) into electricity than conventional silicon solar cells can. Such a breakthrough could lead to high-efficiency cells, with layers of different materials, each optimized to capture a different part of the solar spectrum. Efficiencies greater than 40 percent have been recorded for these "multijunction" types of cells, but they are still laboratory prototypes.

We also need to capitalize on the knowledge already acquired from years of laboratory research. We need to connect the disciplines and sectors (optics, electronics, flat screens, and LED), transfer their results to a large scale, and integrate them to achieve innovations in the electrical, electronic, and optical configurations of solar cells and other solar devices. The goal is to reduce material consumption and increase light-absorbing and energy-transforming capability.

Research throughout the world is promising. Thin films based on minimal amounts of silicon and other materials are expected to make a significant impact on the market. In addition to using less photovoltaic material, thin films can be deposited on large surfaces, leading to the expectation of a significant cost advantage. However, the efficiency of thin films of this type is even lower than that of traditional cells, and for the time being, the cost advantage has not proven out. Cadmium telluride is an exception, but it has material availability limitations and potential toxicity problems.

Other promising avenues use low-cost or very innovative materials, such as polymers to produce plastic cells or nanostructure materials like quantum dots or quantum wells with tunable light-absorbing properties. Nanotubes, nanothreads, and carbon nanospheres are still largely unexplored in terms of light absorption or the transport of electrical current in solar cells.

It is now likely that in a reasonable amount of time, say, by 2020, these areas of research could produce a breakthrough capable of changing the terms of the solar equation. Even so, it will take more time to develop commercially, so it is unlikely that by 2030 solar energy will come out of the obscurity to which the numbers have relegated it.

This does not diminish the basic hope that I expressed at the beginning. In terms of energy potential, solar radiation is the only renewable alternative source capable of supplying our needs. It is a gift of nature, and taking advantage of it depends only on human intelligence. For this reason, we must finance advanced research to overcome the obstacles that seem insurmountable today. In particular, interdisciplinary research should yield the kinds of unexpected results we need. The convergence of quantum physics, biochemistry, molecular biology, electronic engineering, and other disciplines can give us answers to unresolved questions, offering new systems capable of cheaply exploiting this free, inexhaustible, and clean source on a massive scale.

It is true that all this will take a long time, but once it starts, it will be the real energy revolution of the twenty-first century.

Statistics on Solar Power

Table 9.1. Installed solar power electric generation capacity (Top 10 countries and world total, 2008)

Country	*Installed capacity (MW)*
Germany	5,308
Spain	3,223
Japan	2,149
United States	1,173
Italy	443
South Korea	350
China	145
Australia	105
India	90
France	87
World total	**14,730**

Table 9.2. Solar power electricity production (Top 10 countries and world total, 2008)

Country	*Production (GWh)*
Germany	4,000
Spain	2,492
United States	933
Korea	264
Italy	193
China	121
Australia	92
India	79
Netherlands	50
Belgium	40
World total	**8,900**

CHAPTER 10

Geothermal Power: The Primitive Energy of the Earth

Geothermal energy is the only renewable source that lacks a connection, direct or indirect, to the energy of the sun. Hydroelectric power, wind power, and biomass are tied to the sun, as is solar power, of course.

Like the other renewable sources, using geothermal energy is not a new idea. The heat brought to the surface in hot water springs was used for Roman baths and other uses in past civilizations. Its first industrial use goes back only to the beginning of the nineteenth century. At that time, French entrepreneur François Jacques de Larderel set up a system to collect and exploit the steam from geothermal sources to extract boric acid in the area around the town of Pomarance, Tuscany, which later took his name: Larderello.

In 1904, the first experiments generating electricity from geothermal sources also took place in Larderello. Piero Ginori Conti used an alternating engine fed by steam from the Earth to move a dynamo. In 1913, the first proper geothermal electrical power plant was built in Larderello, equipped with a 250-kilowatt turbine. From these first experiments in Italy, the sector developed throughout the twentieth century, but with many limitations.

Today, geothermal power has a negligible impact on the world mix of energy sources. Nonetheless, geothermal utilization has had a big impact in a few locations where high grade natural resources are found, such as Iceland, New Zealand, and several Central American countries. Although it remains an interesting resource with some prospects for development, it will be difficult for energy from the Earth's core to assume an important role in satisfying world energy needs for electricity, at least in the near term.

The Earth as a Source of Energy

The ancients understood that there were sources of heat under the Earth's surface. They based their knowledge on the observation of various phenomena: volcanic eruptions, hot water springs, fumaroles, and such. Although they sometimes exploited the benefits of this underground energy, our ancestors did not understand its causes or the phenomena that brought it to the surface. Instead, they often associated the phenomena with gods and demons. The ancients used geothermal energy when natural conditions were right, when it sprang forth naturally to the surface, without pushing beyond, much as they did petroleum.

Today, we know that the interior of the Earth provides an energy flow—partly from the decay of radioactive elements (including uranium), partly from the thermal conduction of heat through the Earth's crust as a result of the very slow cooling of the most remote regions of our planet—going back to its very birth.[1] For this reason, the temperature of the Earth increases noticeably with depth, on average 25–30°C per kilometer (72–87°F per mile).[2] This number can have significant local variation, and it can be modified by special conditions that cause temperature inversions. It tends to increase near the edges of the Earth's tectonic plates that comprise the crust of the Earth and in areas where the crust is stretched more thinly, where high-temperature magma can approach the surface. These are fertile areas for seismic and volcanic phenomena.

In reality, the presence of a subsurface source of heat is only one of the elements necessary for our technology to exploit a geothermal resource. The source must coexist with a *reservoir* of very permeable porous rock formations (like the reservoirs containing petroleum and gas) and trapped water or steam. All geothermal systems in use today contain these three ingredients in their natural state. These systems are commonly referred to as *hydrothermal reservoirs*. In most geologic settings, one or more of these ingredients is missing. If geothermal energy is to become a global resource, we will need ways of enhancing or engineering permeability in hot rock where it does not exist naturally.

Fractures in the rock above a reservoir of this type can feed hot springs, geysers, and fumaroles, which can easily be used directly, as the ancients did. However, exploiting geothermal sources on a significant scale cannot be limited to collecting natural surface heat flows. As with oil and natural gas, the subsoil must be perforated for direct access to the deposits that contain the geothermal fluid, and there must be enough down there to justify the investment.

The depth of the wells and the energy quality of the fluid depend on the local geology. There are places where magmatic rock comes close to the surface, making it possible to obtain high-energy steam from wells less than 2 miles (3 kilometers) deep. Examples include the fields at Larderello and The Geysers in California, about 70 miles (100 kilometers) north of San Francisco. Today, The Geysers has the largest collection of geothermal power plants in the world, with a combined installed capacity of 750 MW.

Areas on the planet where these conditions occur are relatively rare. In most places, the geothermal fluid is a mix of steam and water or simply hot water, if the reservoir is close to the surface. If a natural steam or hot water deposit is exploited intensively, reinjection wells must be drilled upstream from the source to set up a recharging cycle.[3] Water is reinjected into the field, and the reservoir acts like a boiler, transforming the water into steam, which then comes out at the surface, and the cycle is repeated.

Geothermal Energy Today

Currently, heat from the Earth is used indirectly to generate electricity and directly for residential and industrial heating. Geothermal steam of at least 150°C (300°F), after separating it from the water, can be sent directly to a turbine driving an electric generator. The simplest installations, suitable for many small-scale applications up to 5 MW, let the steam expand through the turbine until the steam reaches ambient pressure, then release it into the atmosphere.

Larger, more complex power generation installations fed by several wells are more efficient and can have capacities exceeding 100 MW. More energy is wrung out of the steam by expanding it through the turbine to below atmospheric pressure in a condenser, then pumping the condensate back below the surface to recharge the reservoir and eliminate liquid effluent. These are mature technologies, widely tested and available on the market.

For some decades now, another kind of electrical power plant has been available, capable of exploiting temperatures even below

100°C (212°F) on a small scale (a few megawatts). In these *binary cycle systems,* the geothermal fluid heats a pure liquid (like water) in a boiler, transforming it into hot steam. The steam is expanded, then condensed into a liquid, and vaporized again. This closed cycle is similar to the traditional fossil-fuel fired or nuclear steam power plant. After giving up its heat, the geothermal fluid can be reinjected into the subsurface. Using relatively low-temperature sources severely limits the yield of these types of installations, with efficiencies of 5–20 percent, depending on the type of primary source and the technology used.

At present, the electrical generating capacity from geothermal sources throughout the world totals over 10 GW. Half of it is concentrated in only two countries, the United States (2.9 GW) and the Philippines (2.0 GW).[4] Electricity production is about 57 TWh, only 0.3 percent of the kilowatt-hours consumed worldwide, although it is an important and cost-effective source in some cases.

Medium- and low-temperature geothermal heat (from 30°C [85°F] to 150°C [300°F]) is used directly for many uses: home heating, pools, thermal spas, greenhouses, aquaculture farms, and manufacturing processes that require drying, sterilization, or distillation.[5] The total installed thermal power of these types of installations is about 17 GW, producing yearly the energy equivalent of around 400 million barrels of oil.[6] Between heat and electrical power, geothermal energy supplies about 0.4 percent of global primary energy.[7]

Geothermal power is also associated with low-temperature climate control, made possible by the relatively constant temperature of the ground, water from lakes and rivers, or surface water tables. This sector uses thermal machines called *geothermal* or *ground heat pumps,* similar to the conventional air-to-air heat pumps of air conditioners, but more efficient.

During the winter heating season, geothermal heat pumps absorb heat from the ground at low temperature and return it at a higher temperature,[8] a characteristic that is used in the winter to heat buildings and homes.[9] In the summer, the process is reversed. Even taking into account the electricity used to operate them, geothermal heat pumps can save significant amounts of energy over conventional heating and cooling systems, so that in recent years they have succeeded well on the market. Between 2000 and 2007, heat pump capacity grew 260 percent worldwide. The fastest growing market was in the United States, with more than 800,000 geothermal heat pumps installed. Next came Sweden where some of the most important heat pump manufacturers are located (270,000 heat pumps by the end of 2006, 45 percent of the EU market). In June 2009, U.S. energy secretary Steven Chu announced that nearly $50 million from the American Reinvestment and Recovery Act would be aimed at speeding up the commercial deployment of geothermal heat pumps.

The Limitations

There are essentially two reasons why the contribution of geothermal energy to the energy consumption of the world is so small: the availability of the resource and its cost compared to other conventional energy sources.

Concerning availability, except for the specific case of heat pumps, the geothermal resources that can be exploited with current technology are naturally occurring hydrothermal systems concentrated in a few areas of the planet. In addition to being relatively close to the surface, nature has provided sufficient formation permeability and a large source of contained hot water or steam. The hydrothermal areas most exploited for electricity generation are the central part of the Pacific coast of the Americas (especially California, Mexico, Costa Rica, and El Salvador), the Philippines, Indonesia, Japan, central Italy, New Zealand, and Iceland. There is an additional problem, however: often where these resources are available, there is no consumption market nearby, a fact that especially penalizes thermal applications.[10] Even in areas where it should be theoretically available, the economic feasibility of geothermal energy can be undone by cost factors, from the depth of the deposits to the energy content and productivity of the fluids that are extracted.

The depth of the deposit is certainly the most significant of these factors. Drilling costs increase exponentially as we push below the surface, much more than they do with oil. In general, the cost of geothermal wells is two to four times higher than that of typical oil wells at depths of 4 km or less. However, their cost increases at a slower rate than oil and gas wells, and crossover occurs between 5 km and 5.5 km.

Concerning the energy content and the type of extracted fluid, its value increases with its temperature and/or its steam content. These determine possible uses, which naturally have to be matched with local market needs. The composition of the geothermal fluid must also be considered, because the presence of dissolved salts or gas (e.g., ammonia or hydrogen sulfide) may require the use of special materials or costly treatment systems.

Finally, the productivity of the reservoir determines the scope of the initiative and its longevity. As with a petroleum deposit, a less productive geothermal deposit may prove uneconomical to exploit, even at a shallow depth.

All these elements (along with financial factors, such as equity and interest rates on invested and borrowed capital) determine the cost per kilowatt-hour of geothermally produced electric power, a cost that appears quite variable. Under the best conditions, when high-temperature steam or hot water is available (e.g., Iceland, Larderello in Italy, or The Geysers in California), costs can be very low (even 5 cents per kilowatt-hour for new construction) with very low operating costs (e.g., 1.5–2.5 cents/kWh at The Geysers).[11] However, the best conditions are rare, limited to shallow deposits of high-energy fluids near consumer markets or where sufficient electricity transmission capacity is available. Elsewhere, the costs go up, and if the resource cannot even be used for heat, it becomes unusable.

The challenges for geothermal energy do not end here.

Exploiting geothermal sources causes perturbations in the subsurface thermal profile, especially if a sizable amount of energy is extracted. Repeated extraction over time can modify thermal conditions at depth, making the source less productive unless the withdrawals are slowed dramatically.

Geothermal extraction can also induce seismic phenomena or subsidence and release dangerous substances into the environment from the fluids extracted. With over 100 years of operating experience, much has been learned about the extent to which these effects can be managed. Most commercial geothermal plants are located in seismically active regions where careful seismic monitoring is carried out. Countermeasures to minimize undesirable effects have been implemented, for example, reinjecting the fluids at controlled pressures and rates after using them and using abatement systems that can separate and segregate potential emissions before they escape the plant. In addition, in arid regions, where the availability of cooling water is limited, dry cooling systems have been successfully deployed. All these measures increase the cost.

Geothermal electric power plants that require condensation of the geothermal fluid also have the problem of the visual impact of their cooling towers. The towers release a large amount of low-temperature waste steam from the condensation process.

Finally, the geothermal resources used so far (except for heat pumps) have only been of the *hydrothermal* type, which involve water or steam naturally bringing the heat of the subsoil up to the surface.

All this undermines the statement that, on paper, geothermal sources could satisfy many of the energy needs of the planet for a long time.[12] As with other renewables and unconventional fossil resources there is a chasm between the theoretical potential and the actual possibilities. Unfortunately, a large part of the geothermal energy is simply too expensive to be exploited.

A Challenging Future

Considering the limitations, it is probable that in the coming decades the role of geothermal energy will remain very constrained, at least until 2030. Its future depends essentially on the development of technical capability to extract energy from geological structures other than the ones used so far. At this time, there are three areas of major interest: geopressurized systems, magmatic fluids, and hot dry rocks (HDR) or, as they are now commonly called, enhanced or engineered geothermal systems (EGS).

Geopressurized systems consist of porous rocks containing hot water under pressure (and often methane), trapped between impermeable rocky strata. Resources of this type are sizable, but research so far has not led to commercial applications because of the relatively high cost of development.

Using *magmatic fluids* involves exploiting the heat from magmatic intrusions close to the surface (within 6 miles [10 kilometers], considered today to be the drilling limit). This option poses a very complex technical problem, because it involves operations at great depth and in very hot environments (many hundreds of degrees).

The path that could lead geothermal power to assume a more important role in the field of energy is the exploitation of hot dry rocks through EGS, which constitute the most widespread type of resource. They consist of rocky strata at relatively high temperatures to depths of 10 km or less, but without sufficient permeability, porosity, and/or presence of subterranean water.[13] An important study by the

Massachusetts Institute of Technology estimated that a research and development investment between $800 million and $1 billion over a fifteen-year period could enable the commercial deployment of EGS to exploit HDR. This would allow for the installation of more than 100 GW of new geothermal power capacity in the United States alone by 2050.[14]

Geologic structures of this type are available everywhere, and their advantage is that they offer temperatures of 300°F (150°C) at relatively modest depths (about 3 miles [5 kilometers]). The problem with HDR is usually that the rocks are not porous enough to allow the passage of water or contain no water at all. Artificial systems would be needed to overcome these limitations. These systems exist on paper, using tools that have already been tested and adopted in the oil sector. For example, hydraulic, chemical, or thermal methods could create fractures in the underground rocks perpendicular to a first well. A second well would then be drilled to inject water into the fractured structure (water may also be used to fracture the rock), setting up an artificial circuit to bring the heated water to the surface, where it flashes into steam and it is used to generate electricity. When it returns into its liquid state, the water is re-injected into the ground.

It is easy to see the high potential of this type of solution. Research and experimentation in this field have been going on for some decades, with many aspects of technical feasibility demonstrated successfully. Nevertheless, for the past 15 years there has been a relatively low level of R&D funding available, inadequate to sustain commercial-scale EGS field demonstration projects in the United States and Europe.[15] Thus, development risks for ESG are presumed to be high and the prospect of large-scale industrial application still seems distant. The principal problems are the high cost of drilling several wells, and the need to show sufficient durable productivity, while controlling the losses of circulating water by dispersion.

These difficulties can be overcome, but it will take time to resolve them cost-effectively. Perhaps a contribution to the development of geothermal energy could come from focused public incentives.[16] Unlike other renewable sources of energy such as solar, biomass, and wind, geothermal energy has always been overlooked by governments, placing an additional obstacle in its path to develop improved technology, reduce costs, and expand global deployment.

Statistics on Installed Geothermal Electric Generation Capacity

Table 10.1. Installed geothermal electric generation capacity (Top 10 countries and world total, 2008)

Country	*Installed capacity* (GW)*
United States	2.9
Philippines	2.0
Indonesia	1.2
Mexico	1.0
Italy	0.8
Japan	0.5
New Zealand	0.5
Iceland	0.5
El Salvador	0.2
Costa Rica	0.2
World total	**10.2**

* Installed geothermal capacity is greater than geothermal electricity production, which totals 9.1 GW worldwide.

CHAPTER 11

The Hype about Hydrogen, the Hope about Electric Cars

Dealing with hydrogen in a book about primary energy sources is an anomaly, because hydrogen is not a primary energy source. It is an *energy vector*, like electricity, meaning that it is obtained by using energy from another source. Hydrogen is the most abundant element in the universe. However, it is not found on Earth in any readily usable form, but only in combination with other elements forming familiar compounds such as water, petroleum, and natural gas. This means that to obtain pure hydrogen (H_2) and to use it as an energy source, it must be extracted from the compounds in which it is found. This makes hydrogen an energy vector.

Nevertheless, dealing with hydrogen is almost inevitable. In recent years, a deluge of articles, books, and enthusiastic statements from futurists and self-styled experts have forecast the advent of an energy revolution based on hydrogen, so it would seem suspicious if I were to omit hydrogen. Apart from the semantics of sources and vectors, the revolution painted so simplistically is a long way from reality, because the problems with producing, transporting, storing, and distributing hydrogen at economically acceptable levels are still insurmountable. This is particularly true if we are looking for an economy based on zero-emissions hydrogen. It is more realistic, then, to recognize that electricity rather than hydrogen should power our cars in the future. But an electric solution must also overcome some significant obstacles.

The Problems, Upstream and Down

More than 95 percent of the hydrogen used in the world today is produced from natural gas (a little less than 50 percent), petroleum (30 percent), or coal (just under 20 percent). These are precisely the fossil sources that hydrogen is supposed to displace. This apparent paradox has a simple explanation. Extracting it from fossil fuels is by far the most economical way to do so, even if pure hydrogen remains an expensive product.[1]

In absolute terms, the most economical hydrogen is produced from natural gas and steam using a chemical process called *steam reforming*. This requires much energy and allows conversion of only part of the hydrogen (typically 70 percent) that enters the extraction process as natural gas and steam.

The number-one producer and user of hydrogen in the world (80 percent of total consumption) is the petroleum industry, followed by the chemicals and petrochemicals industries. The former is the industry that conspiracy theorists believe wants to stop hydrogen development on a global scale. In fact, the opposite is true.

The petroleum and petrochemical industries have a vital need for hydrogen for many of their manufacturing processes, from the elimination of sulfur in fuels to improving the quality of heavy crude oils, from the production of ammonia (the basis of many fertilizers) to the production of fuel additives and other products. Therefore, if anyone is interested in producing large quantities of hydrogen ever more cheaply, it is the petroleum industry in particular. Unfortunately, there are objective limits to this goal, limitations that go beyond the production phase. The real culprits include the high cost of the entire hydrogen chain, including transportation, storage, and distribution. In large part, the problems stem from the very nature of this gas.

Hydrogen is a fuel with a very high energy density by weight, but with low energy density by volume. At ambient temperature and atmospheric pressure, the energy in 1 kilogram of hydrogen occupies more than 12 cubic meters. This volume is ten times greater than that occupied under the same conditions by a kilogram of natural gas. Consequently, to transport significant amounts of energy by tanker truck or gas pipeline requires high pressure (200–300 bars), much higher than what is required to transport natural gas (typically about 75 bars).

Naturally, if the cost of transporting hydrogen in gaseous form is high, transporting it in liquid form (like methane) is even more expensive. Liquefying hydrogen requires a complex and burdensome process of cooling the gas nearly to absolute zero (−253°C for hydrogen, compared to −161°C for methane). This process costs as much as the steam reforming, doubling the total cost. Transportation in special cryogenic tanker trucks and storage near the point of use add to the costs, as do the installations for regasification and distribution.[2]

Therefore, production and transportation costs make hydrogen far more expensive than gasoline or diesel. When the cost of oil is about $60 per barrel, the cost of producing and transporting 1 kilogram of hydrogen (from natural gas, the most economical source) is about $4–$5, while that of traditional fuels is about 55–60 cents per kilogram.[3]

It should be immediately pointed out that this type of comparison is deceiving, because—given its high energy density by weight—1 kilogram of hydrogen releases three times the amount of energy of a kilogram of gasoline (almost 30,000 kcal per kilogram compared to 10,000 kcal per kilogram). To be done accurately, comparisons between hydrogen and gasoline or diesel must be made in terms of energy equivalence, not weight or volume.

The cost of producing and transporting a quantity of hydrogen from natural gas with the same energy content as a given volume of gasoline is about three times as much. This inequality does not even take into account the cost of distributing hydrogen to the final customer or of the dedicated equipment for its use and storage, which are prohibitive today.

Problems Using Hydrogen for Transportation

Let us consider the most promising and efficient use of hydrogen for transportation: fuel-cell technology. Fuel cells already cost between twenty and eighty times more than internal combustion engines (about $30 per kilowatt of installed power compared to $500–$2,500 for fuel cells).[4]

In addition, storing hydrogen in the vehicles is complicated and expensive, whether as a liquid (pressurized at cryogenic temperatures) or as a gas (very high pressures). In the latter case, storage pressures up to 700 bars are required because hydrogen has a very low energy density per cubic meter. This means that a tank with a capacity of 5 kilograms of hydrogen, enough to travel 300–350 kilometers (around 200 miles), takes up ten times more space than the volume of gasoline needed to run the same distance.[5] More important, its cost is much higher, about $3,000 to $4,000.

To understand the difficulties in this sector, consider the BMW Hydrogen 7. BMW announced the launch of 100 street prototypes of this car, with a hydrogen-fueled internal combustion engine (strictly traditional technology) and a dual fuel system, which would have a range of 200 kilometers (125 miles) on hydrogen and 500 kilometers (300 miles) on gasoline. After a grand advertising campaign in the autumn of 2006, the plan seemed to stop suddenly, probably because of the prohibitive cost of the vehicles and the difficulty setting up even a limited infrastructure for hydrogen refueling. BMW had announced that it wanted to lease the vehicles to a select clientele who could afford lease payments of several thousand dollars per month. The technical choice of an internal combustion engine was also severely criticized, being the least efficient fuel cell solution.

Some automobile producers continue to trot out hydrogen cars during special occasions to promote their commitment to a sustainable future. Those cars are not for sale, however, because they would probably cost more than $300,000 each. They belong to small fleets of a few dozen units owned and operated by their manufacturers for public relations purposes.

Storing hydrogen gas constitutes one of the biggest limitations to developing a hydrogen automobile sector (or for transportable energy in general). Neither pressurization nor cryogenic transport is satisfactory in terms of cost and inconvenience.

Research to overcome this problem using chemical and chemical-physical methods is under way. There are chemical compounds and porous systems that behave like sponges and could hold large quantities of hydrogen in reduced volumes. To be useful, these systems need to meet three fundamental prerequisites:

1. Be able to release hydrogen easily on demand
2. Have simple, inexpensive development processes
3. Be safe (neither explosive nor flammable nor toxic)

Traditional research in this field has concentrated on special compounds called *hydrides*, which release hydrogen in the presence of water. The cost and the properties of hydrides have not yet yielded a satisfactory product.[6] Instead, there is hope in trapping hydrogen in nanostructure materials based on carbon. These would be proper sponges, from which we could extract hydrogen safely and without losses. To this end, researchers have studied nanotubes, nanofibers, and more complex structures,[7] but much more research work will be needed to develop products engineered with all the necessary characteristics to use them in the transportation sector.

In any case, to establish themselves on the market, fuel-cell automobiles will need to beat their competitors. Among these, hybrid cars are certainly establishing themselves on the motoring scene with a mature and more promising technology. Internal combustion engines continue to evolve, and they have the advantage of a capillary network for fuel distribution.

This last aspect constitutes a formidable obstacle for hydrogen to penetrate the market as a transportation fuel. According to estimates by the Argonne National Laboratory, building a hydrogen distribution network for the United States could cost as much as $500 billion to supply 40 percent of the light vehicles in the country. This poses the classic problem of sequencing investments.[8] On one hand, it will be difficult to develop an expensive infrastructure for hydrogen distribution without an adequate fleet of hydrogen cars. On the other hand, no one will produce a large number of hydrogen vehicles unless there is a significant distribution network.

Thus, it does not come as a surprise that in May 2009, the Obama administration slashed the U.S. Department of Energy Fuel Cell Program by $100 million, considering that the need for better fuel cells and the lack of infrastructure would preclude a transition to a hydrogen car economy anytime soon.[9] Most of the remaining budget ($68 million) of the program will be devoted to applications in buildings rather than in cars.

The last problem of using hydrogen on a large scale is safety. Even today, there is a *Hindenburg syndrome* associated with this gas, so called in reference to the German dirigible *Hindenburg* filled with 200,000 cubic meters of hydrogen, which exploded while docking at Lakehurst, New Jersey, in 1937. According to most studies, the real cause of the accident was not the hydrogen but the covering of the dirigible, which was built of a very flammable compound.[10] The fact remains that some of the characteristics of hydrogen require specific measures to use it safely. These include its great flammability and volatility, its low flashpoint, the fact that it is odorless and colorless, the fact that its flame is invisible during the day, and the speedy propagation of the flame itself. In other ways, hydrogen is altogether similar to traditional fuels (gasoline, propane, natural gas) in terms of potential danger.

The Future

If hydrogen could really assure zero pollution, the subject of its cost could be seen in a completely different light, but this prospect is not realistic given the state of technology today. Obviously, producing it from natural gas, oil, or coal does not avoid the collateral environmental effects of using fossil fuels, only adding the energy cost to the cost of production.

Some potentially interesting alternative technologies involve converting biomass into hydrogen using thermal-chemical gasification or biological processes based on microalgae. Many demonstration initiatives are under way, trying to eliminate uncertainties about the cost and handling of the by-products of gasification. Biological processes, although promising, are still at the research stage.

The only widely available hydrogen source holding the promise of a completely clean fuel and zero emissions is water. The problem is how to extract the hydrogen that it contains. We can already do this today using electrolysis, but this uses great amounts of electrical energy, which, in turn, is most economically obtained by burning coal, oil, or natural gas. This vicious circle is aggravated by the fact that carbon dioxide emissions from the electrolytic production of hydrogen (when using fossil fuel for electricity) to its final use in a fuel cell vehicle would be the same as those from a modern gasoline-powered vehicle, 20 percent more than the latest-generation diesel engine, and 100 percent more than a hybrid vehicle (internal combustion engine plus electric motor) using natural gas.

In summary, it is true that the reaction between hydrogen and oxygen (as in a fuel cell) produces only water and no harmful exhaust. The problem is upstream, in the way that the electrical energy is obtained to extract the hydrogen, which remains expensive and polluting.

The only real alternative would be to produce electrical energy for electrolysis using nonpolluting renewable sources, for example, solar energy (a process known as hydrogen photo splitting). Photovoltaic cells could power commercial electrolyzers directly, electrolysis being the most mature technical option.[11]

Nevertheless, the high cost of renewable sources in general makes it very expensive to use any of them to generate electricity for electrolysis. Hydrogen from renewables would cost five to thirty times more than gasoline or diesel for the same energy content.[12] The hope is that progress in renewable technologies, especially solar energy, can change the terms of the issue dramatically.

There is still the possibility of using nuclear energy to power the electrolysis of hydrogen from water or to obtain it using thermochemical techniques and high-temperature reactors.[13] These options would have zero-emissions outcomes at a lower cost.[14] Both options require ample nuclear energy, however, which is not likely under the present state of affairs.

Thus, the prospect of a hydrogen economy is far from happening—very far. Explaining the U.S. government decision to cut financing for the hydrogen program, Secretary Chu said: "We asked ourselves: is it likely in the next 10 or 15, 20 years that we will convert to a hydrogen car economy? The answer, we felt, was *No*." Enormous efforts in scientific research will be needed to make the promises of this widely available gas less uncertain. This research must move in parallel with efforts to give a serious role to renewable energies, which are required for true zero-emission hydrogen. And it must work to lower the cost of storing and distributing hydrogen and the costs of the systems that use it most efficiently, such as fuel cells. In this last regard, the prospects are encouraging. However, they are not so encouraging as to justify the predictions of the harbingers of a hydrogen economy.

The Electric Car: Old Utopia or Future Reality?

What appears more viable than hydrogen-fueled cars would be the widespread diffusion of electric cars, of which the hybrid cars of today could be the precursor. Yet, even this perspective must be carefully analyzed.

Only car historians still remember that the electric car is not a real novelty, but a dream that went bust early in the twentieth century. At that time, it spread more rapidly than the gasoline car and for a while, it epitomized the future of civil transportation.

But the dream had a short life. In 1908, Henry Ford rolled out his gasoline-powered Model T, the first mass-produced car, so inexpensive that any worker could afford it. During the same period, the price of oil and gasoline decreased noticeably. In 1913, the discovery of thermal cracking made it possible to obtain much more gasoline from a single barrel of oil.[15] Finally, the electric car could never overcome another major limitation: its range was only 50 to 80 miles (80–130 kilometers) on a battery charge, and battery chargers were very rare. Above all, as James Flink wrote, the electric car "was far more expensive than the gasoline car to manufacture and about three times more expensive to operate."[16]

The many attempts during the twentieth century to bring back the electric car always faced three major problems. First, during the last century the price of oil stayed low consistently, except for short-lived spikes. Second, it was not cost-effective to develop a car that was much more expensive and had a much more limited range than a gasoline or diesel car. Third, we have never developed batteries capable of storing large amounts of energy.

Now the dream of the electric car is back, through the great expectations surrounding *plug-in hybrids*. Models are being developed almost everywhere in the world, and the Obama administration is particularly keen to promote this technology. It has set a goal of having a million plug-in hybrids on the roads by 2015. However, plug-in hybrids are still plagued by some of the problems that condemned the electric car to oblivion in the past.

The novelty of the plug-in hybrid lies in the fact that the battery can be recharged through a normal electrical outlet. The first large-scale production model, the Chevrolet Volt, is expected to reach the U.S. market toward the end of 2010. It is a *series hybrid*: the car is also equipped with a gasoline engine (hence the term *hybrid*), which is activated when the battery is discharged (that is, after about 40 miles [65 kilometers]). However, the purpose of the gasoline engine is only to recharge the battery, not to propel the vehicle. This distinguishes series hybrids from the hybrid cars already on the market, which are *parallel hybrids*, like the best-selling Toyota Prius. The parallel hybrid is powered by a gasoline engine and uses the more efficient electric engine only as an auxiliary source of power, particularly in stop-and-go urban traffic. The series hybrid, in which the internal combustion engine is ancillary, is the one that could really open the way to battery-only electric vehicles.

The first problem with series hybrids is the absence of a range of affordable models; only very expensive vehicles are currently on the market. Developing reliable, high-capacity lithium-ion batteries (the kind of rechargeable batteries in cell phones) will be crucial to the future of this technology.[17]

Cost will be another essential issue. The estimated cost of the Chevrolet Volt, at least $40,000, exceeds the willingness or ability to pay of most potential customers.

Parallel hybrid cars are much less expensive. The Toyota Prius currently sells for $22,000 for the basic model, a premium of about $5,000 to $8,000 over the equivalent gasoline model. Until now, traditional hybrids have been quite successful in the U.S. market, thanks to the high price of oil and public subsidies. In the five years before the oil price collapse of the second half of 2008, the hybrid market share had jumped from 0.5 percent to 2.5 percent. Then growth suddenly stopped.

This indicates that the cost factor will be crucial for the commercial success of series hybrids. It is still true that they consume a very small amount of gasoline, whatever it takes for a partial battery recharge, so that the cost per mile is very low (even lower than for parallel hybrids). However, if the purchase price of the vehicle is too high, future savings will be unlikely to persuade consumers to buy the car. For this reason, the $7,500 tax credit in the Obama plan for every buyer of an electric car will probably be inadequate, unless foreign producers enter the U.S. market—first and foremost the Chinese, who seem to be already capable of producing electric cars at affordable cost.

An example is the BYD Group, an automaker and lithium-ion battery producer based in the south of China and guided by rising entrepreneur Wang Chuan-fu. The group already manufactures and sells a plug-in series hybrid, the F3DM. Its reliability has yet to be put to the test, but the legendary investor Warren Buffett recently put a bet on the venture, acquiring a 10 percent stake in it. The F3DM car is sold on the Chinese market for the equivalent of $22,000, and Wang has stated his intention to sell it for that price on the U.S. market, too. This opens up a very sensitive issue.

President Obama is trying to create jobs in the United States, while at the same time saving U.S. car giants from bankruptcy. The first goal could be pursued even if a foreign (i.e., Chinese) company were to produce cars in the United States, but this would not help

GM, Ford, and Chrysler out of their trouble. Protectionist attitudes remain strong in the U.S. Congress. In fact, in approving the initial aid package recently, Congress limited its applicability to firms that have been operating on U.S. territory for at least twenty years, effectively ruling out most foreign producers. Moreover, if Wang and other Chinese entrepreneurs were to produce their cars in the United States, they could not keep their labor and manufacturing costs at the low levels they enjoy in their homeland.

To promote the birth of an American electric car industry, the president and Congress will probably need to increase the subsidies to at least $10,000 for each car purchased, according to one Cambridge Energy Research Associates study,[18] for the first million cars by 2015. This would amount to $10 billion, a relatively modest sum considering the size of the U.S. budget. One million electric cars on the roads of America, with annual sales of 200,000, would represent a tiny share of the 240 million vehicles in a market where, before the crisis, more than 16 million cars were sold every year. However, that number might represent the critical mass needed to sustain a technological revolution, which could change the very concept of a car—a revolution that would once again be "made in USA."

Nevertheless, this remains an ambitious target, especially if American consumers find the required savings in more efficient, conventional cars. We should always keep in mind that any technology—and the electric car is no exception—never competes only with itself, but also with rival technologies. It is likely that traditional gasoline and diesel still have much to say. The financial crisis of 2009 gave the Obama administration an unprecedented opportunity to shape the future of the auto industry, even as the government acquired a majority stake in General Motors. However, the crisis has also imposed on the administration an urgency to be pragmatic and to pursue the most efficient medium-term options to reach its environmental goals.

The dream of an electric car could still crash without becoming a reality, at least in the West. In the East, this dream could still become a reality, but in a different form.

At the beginning of the twentieth century, Henry Ford had announced his Model T by saying, "It will be so low in price that no man making a good salary will be unable to own one."[19] Today, Tata Motors chairman Ratan Tata hopes to replicate Ford's success, a hundred years later and 8,000 miles from Detroit. In July 2009, he handed the keys of the first Tata Nano, long awaited queen of ultralow-cost cars, to Indian civil servant and Mumbai resident Ashok Vichare for $2,500. In the last three years, small and economical cars, fueled by gasoline, have emerged as an option for the Far East markets. At the same time, sales of two-wheeled electric vehicles have jumped: In China alone last year, 21 million such vehicles were sold, compared with 9.4 million cars. In the future, those two trends could combine to form the low-cost, small electric car. It could give many people access to reasonably priced, comfortable, low-polluting mass motorization, especially in the Asian megacities.

In other words, the electric car may have more chance of success in markets that are still untouched by mass motorization than in the mature ones. Even if the United States made a fundamental contribution to the development of rechargeable batteries, it is in Asia where the manufacturing expertise has matured and the capital investments have been made.[20]

If this is the case, the electric car of the future could come from the East, and from there, when perfected, it could try to conquer the West.

CHAPTER 12

Energy Efficiency: The Invisible Source of Energy

Before coming to conclusions, I must devote a chapter of this book to what I consider the most important and effective source of alternative energy: energy efficiency.

Why do I consider efficiency an energy source? Because it means either using less energy to get the same result (for example, switching from incandescent to compact fluorescent light bulbs, which use one fourth the electricity) or getting more energy from the same amount of raw materials (for instance, getting more electricity from the same amount of coal thanks to a modern coal-fired plant). Essentially, it means producing the same with less.

Moreover, I consider it the *most important* source of alternative energy because it can deliver both

- the best short-term results in terms of reduced consumption and reduced emission of greenhouse gases
- the best cost-benefit ratio in many of its potential applications

However, energy efficiency presents its dark side as well. It is difficult to evaluate its effective potential, and its strongest advocates are prone to blunder into a variety of traps that come between the attractive idea and its implementation. Thus, understanding energy efficiency's pros and cons is necessary to maximize its advantages and avoid reducing it to an empty or even harmful slogan.

Why Energy Efficiency?

The way the world uses energy may be illustrated by the metaphor of a man who, in order to destroy an anthill in a corner of his garden, sets fire to the whole garden. As a species, we tend to use too much energy to perform a given task.

From a thermodynamic point of view, most of the primary energy we use is wasted in the processes that convert it into usable energy for its end uses. In this case, "wasted" does not mean "destroyed," since, by the first law of thermodynamics, energy can neither be created nor destroyed. It simply means that much of the energy we exploit is useless to the task for which we intend it. This inefficiency occurs for three main reasons.

First, by the second law of thermodynamics, the energy content of a given resource always degrades when it is used or transformed in an isolated system.

Second, we waste energy because of the imperfect conversion efficiency of available power plants, transmission grids, motor vehicles, appliances, and all the myriad machines and manufacturing processes that use and transform energy for the needed tasks. For example, we lose part of the energy content of oil as we transform it into gasoline. Then we use only a modest fraction (less than 10 percent) of the gasoline energy to move our cars; the rest is wasted in the process. Similarly, we convert into electricity only a part of the energy content of coal (on average, less than 40 percent). A further fraction of that electricity is then wasted in its transmission and distribution. Finally, we waste most of the electricity that we receive as end consumers to perform tasks that ideally require much less of it. For example, when we switch on an incandescent light bulb for illumination, a large fraction of the electricity we pay for is wasted to produce heat rather than light.

Third, we often waste energy simply because it is so cheap that we do not trouble ourselves to conserve it, as when we leave the lights on when nobody is there.

In 2000, a task force promoted by the United Nations Development Program released its "World Energy Assessment," which estimated the global loss of energy through all stages of its transformation and consumption. According to that assessment, the world's energy efficiency in 2000 was 37.5 percent, meaning that 62.5 percent of primary energy content was being wasted. In quantitative terms, of the

400 exajoules (EJ, where 1 EJ = 10^{18} joules) of primary energy consumed worldwide in 2000, only 300 EJ was turned into secondary energy (electricity, heat, etc.). In turn, these 300 EJ were converted into only 150 EJ of end-use output.[1]

However, this picture represents only a part of the story.

Thermodynamics enables us to calculate maximum energy efficiency as the lowest ratio of energy input to useful output that is theoretically needed for a given device to perform a given task. This means that, in theory, there should exist "perfect devices" capable of operating on the thermodynamic minimum of energy inputs needed to perform their functions.

In the real world, such theoretical devices do not yet exist. However, in each field of activity there are best-available technologies that offer us possibilities to reduce our consumption of energy.

According to the UN energy assessment, if it had used only the best technologies existing in 2000, the world would have needed less than 60 EJ in that year, instead of the 150 EJ that it actually consumed for all uses. In other words, if world energy use in 2000 had been as efficient as was thermodynamically permitted by the best technology then available, the world could have derived the same level of energy services it actually got while using only 15 percent of the primary energy that it actually consumed that year. Nevertheless, this is a calculation that belongs to theory and not, unfortunately, to the real world in which we live.

The Bright Side of Energy Efficiency

Notwithstanding our energy inefficiency, history offers us several examples of how energy efficiency has helped to cool down consumption trends in the world, without endangering economic growth.

In California, for example, per-capita consumption of electricity has remained flat since the mid-1970s,[2] even though per-capita gross domestic product (GDP) has almost doubled in real terms. As a comparison, per-capita electricity consumption in the United States has increased by 50 percent over the same period. This means that if California had preserved its consumption habits of the late 1960s, today it would need an additional 130,000 tons of coal or 670,000 barrels of oil per day to produce its electricity.

Although California has been more diligent in its energy policy than the rest of the country, the United States has achieved some significant results nationwide as well. According to a study released in 2009 by the American Council for an Energy-Efficient Economy (ACEEE), the U.S. demand for energy and power resources grew by only 50 percent between 1970 and 2008, whereas the national economic output tripled over the same period.[3] Based on an analysis of historical data, the ACEEE study stressed that such a significant reduction of energy usage per unit of output occurred despite a "haphazard and often counterproductive approach to energy efficiency and energy policy." In other words, Americans paid little attention to energy efficiency; it was the emergence of new technologies and tools that required less energy that did the job.

By the same token, without the new technologies and energy-efficiency actions undertaken between 1971 and 2005 (e.g., without cooling down the energy intensity ratio existing in 1971), the primary energy consumption of the European Union would be today 50 percent higher than it actually is. This has made efficiency the top energy source of the European Union so far, as the EU Commission recognized in 2008.[4]

Unfortunately, during the last fifty years, emphasis on energy efficiency has risen and fallen with the price of oil, mainly because cheap oil (which was the rule for most of the twentieth century), and cheap fossil fuels in general, fostered the worst of energy consumption habits. Because these resources were cheap, it was quite natural for people to waste them.

The only relevant efforts in terms of energy efficiency were undertaken worldwide because of the oil price shocks of the 1970s. In the United States, for example, it was at this time (1975) that the administration of President Gerald Ford introduced "corporate average fleet economy" (CAFE) rules, imposing a ten-year target of doubling the efficiency of passenger vehicles to 27.5 miles per gallon. With some minor changes, those rules have not been improved upon since.

With the declining oil prices of the early 1980s, and their eventual collapse in 1986, attention paid toward energy efficiency decreased drastically worldwide. In fact, President Ronald Reagan decreased the CAFE requirements to 26 miles per gallon and cut all the programs established by President Jimmy Carter to develop alternative energy sources. But even if not improved, most of the efficiency measures imposed in the 1970s and early 1980s survived and contributed to curbing the rate of increase of energy consumption of the world.

The abandonment of attention to energy efficiency went hand in hand with the low oil prices prevailing from the mid-1980s through the whole of the 1990s. Among other things, that allowed the rise and explosion of sales of sport utility vehicles (SUVs) in the United States, probably one of the most visible symbols of wasteful consumption habits. SUVs were exempted from CAFE standards (they were placed in the category of commercial vehicles). They represented the new love affair of Americans, and since the mid-1990s their average sales have represented more than 50 percent of the total passenger vehicles sold in the United States.

Given this poor record of the last twenty-five years, the potential savings to be gained through energy efficiency are huge. In this regard, one of the most interesting studies about the effective potential of energy efficiency was released in 2009 by McKinsey & Company.[5] The study was quite innovative because it took a comprehensive economic approach by modeling "deployment of 675 energy saving measures to select those with the lowest total cost of ownership" and ensuring that those measures would have a positive net present value over a ten-year period.

According to McKinsey, a "holistic" U.S. approach to energy efficiency could slash energy consumption in 2020 by 23 percent of the energy demand projected by the U.S. Department of Energy for that year in its business-as-usual scenario (2008). This means a consumption by U.S. end users of 30.8 quadrillion BTUs in 2020, against a projected consumption of 39.9 quadrillion. Such a saving would also avoid the production of up to 1.1 gigatons (billion tons) of greenhouse gases annually, the equivalent of taking the entire U.S. fleet of passenger vehicles and light trucks off the roads.

From an economical point of view, the measures studied by McKinsey would save $1.2 trillion by 2020, while requiring an investment of $520 billion (not including program costs) over a ten-year period. The study suggested that this huge investment could be recovered by a slight increase in the cost of energy to consumers, such as "a system-benefit charge on energy on the order of $0.0059 per [kilowatt-hour] of electricity and $1.12 per [million BTUs] of other fuels over 10 years."[6] The study also stated that introducing a carbon price of $50 per ton of CO_2 emissions would increase savings by an additional 13 percent.

However, the McKinsey study recognized that its "holistic plan" was not easy to carry out, because of its complexity and challenging nature—and also for political reasons. It is worth mentioning that the study considered those savings that could be obtained from energy uses only and did not try to calculate further savings from more efficient systems of transforming primary sources. These could significantly increase the potential for energy efficiency gains in the United States.

The situation does not change if we extend our analysis from a single country to entire sectors of the world's economy. Power generation, for example, absorbs 35 percent of the global consumption of fossil fuels.[7] As we saw in chapter 3, today we must take advantage of combined-cycle gas turbines whose efficiency is close to 60 percent. If we add to this ratio the heat that may be used through cogeneration, efficiency approaches 80 percent. Yet, only 7 percent of global power production originates today from cogenerative combined-cycle gas plants, and the world average efficiency of power plants is less than 35 percent.

Furthermore, the industrial sector has vast room for energy efficiency, even though it has been much more attuned to the subject for economic reasons (e.g., to reduce the amount of energy used per unit of output). It is worth recalling that electric motors used in many industrial processes account for more than 30 percent of all end-use electricity consumed worldwide.[8] By replacing old electric motors with new ones it would be possible to slash electricity consumption by half of the current global production of electricity from nuclear energy.[9] What's more, the higher cost of an advanced motor could be recovered in only two to four years.

Then there is the transportation sector, which accounts for more than half of the global demand for oil products. Technology has brought about dramatic improvements in the energy performance of all forms of transportation. Unfortunately, in most cases, the technological innovations have been used to make bigger and more powerful vehicles (such as SUVs) affordable rather than to cut vehicle fuel consumption.

In 2007, for example, Americans consumed about 26 barrels of oil per year per person (75 percent of which was for transportation purposes), compared to 12 barrels consumed by a Western European. Yet, the standard of living of both areas is substantially similar—and Europe is not a particularly virtuous example of energy efficiency.

Nonetheless, Europe's car fleet consumed on average 37 percent less fuel per mile than America's.

If the average gasoline mileage efficiency of the U.S. car fleet were to match that of Europe (and the U.S. population remained unchanged), the country could save about 4 million barrels of oil per day, or a fifth of its total annual consumption—a figure equivalent to almost the entire production of Iran, the world's fourth largest producer. This would mean a savings of around $44 billion per year, assuming an average oil price of $30 per barrel (a $10 increase of the barrel price would add an extra saving of around $15 billion per year).

To give another view of the importance of such energy savings, 4 million barrels per day of oil may produce annually more than 900 terawatt-hours of electricity—more than the entire nuclear generation capacity of the United States (its 103 nuclear reactors in 2008 produced around 800 terawatt-hours).

Such a savings cannot be obtained in a short time. It needs to be carefully planned, supported by laws, and financed for at least ten years. It takes that long for the automakers to change their models and for consumers to change their vehicles. This is not a revolutionary target.

This reduction in consumption does not even take into account the benefits that would derive from more efficient systems of public transportation, particularly in the big urban and suburban areas of the world. In Hong Kong, for example, public authorities have designed an integrated system of urban transportation that relies mostly on public vehicles. The outcome is a cost to the community that represents 6 percent of Hong Kong's GDP. By contrast, in Houston, Texas, transportation relies predominantly on private vehicles at a cost that equals 12 percent of the city's GDP.[10]

These scattered examples offer an impressive, albeit limited, view of the huge gains that energy efficiency may deliver. These gains tend to increase over time thanks to the unpredictable outcomes of technological innovation, which over history has given us extraordinary new devices, appliances, and machines that have increased our possibilities exponentially while consuming less and less energy. (Think, for example, about computer processors, which have doubled their capacity of processing every two years and today are enormously more powerful, less costly, and yet more energy efficient.)

Some of these new devices are already on our horizon, or already exist and need only to be improved and become less costly. For example, light-emitting diodes (LEDs), in my view, are set to become the future of private and public lighting during the first part of the twenty-first century. Today's white LEDs can last for 50,000 to 100,000 hours of continuous use, making them fifty to a hundred times more long-lived than conventional incandescent bulbs. Also, LEDs last five to ten times longer than compact fluorescent bulbs, which many now see as the new frontier of lighting because they yield an energy savings of more than 60 percent relative to traditional incandescent bulbs. LEDs are even more efficient. They are simply too costly today, and the quality of their light still needs to be improved. But these two obstacles will be overcome with time.

All this said, two questions naturally arise about energy efficiency: Why can't the world be much more energy efficient? Why would a responsible government not rush to implement serious energy efficiency plans? For more than thirty years, energy efficiency guru Amory Lovins has envisaged the possibility of reducing our energy use by 75 percent without adversely affecting our standard of living.[11] Is that really possible?

Unfortunately, calculations based on a holistic approach to energy efficiency have always overestimated the actual likelihood of reducing our energy dependence. Even the most sophisticated econometric models cannot take into account the full complexity of the human, social, economic, and political factors that determine the way we use energy. And most models have usually ignored what I call the "traps" of energy efficiency.

The Traps of Energy Efficiency

For all its magnificent prospects, energy efficiency is beset by many traps that impede full deployment of its theoretical potential.

The first trap is what I call "imperfect substitutability." This is a common mistake of judgment by energy efficiency advocates, who take for granted that some devices are perfect substitutes for other, less energy-efficient ones. In fact, there are at least four circumstances in which substitutability may appear to be an attractive possibility but in fact falls short as a practical economic alternative for consumers and businesses. As adduced by Mark Jaccard, these are:

> First, new technologies usually have a higher chance of premature failure than conventional technologies and therefore pose greater financial risk. New compact fluorescent light bulbs have exhibited higher rates of premature failure than conventional incandescent bulbs, requiring a higher-than-expected financial outlay because of early replacement in some cases. . . .
>
> Second, technologies with longer payback periods (relatively high up-front costs) are riskier if the cumulative probability of failure or accident, or undesired economic conditions, increases over time. . . .
>
> Third, two technologies may appear to provide the same service to the engineer but not to the consumer. . . . Many consumers . . . find compact fluorescent bulbs to be less than perfect substitutes for incandescent bulbs in terms of attractiveness of the bulb, compatibility with fixtures, quality of light and time to reach full intensity. . . .
>
> Fourth, not all firms and households face identical financial costs: acquisition installation and operating costs can vary by location and type of facility. This diversity means that a comparison of single-point estimates of financial costs may exaggerate the benefits of market domination by the highest efficiency technology available.[12]

An additional problem with imperfect substitutability is that it tends to provoke a backlash effect among consumers. Once they have been disillusioned by the poor performance of new and costly energy-efficient tools, they will tend to become much more negative toward *any* new tool promoted as being more energy efficient.

The second, insidious trap of energy efficiency is the "efficiency paradox," also known as the "rebound effect." Historically, greater efficiency in using energy sources has always been followed by an *increase* in the consumption of energy, albeit getting much more output than previously possible.

The subject is well understood and has been extensively studied by economists. As far back as 1865, William Stanley Jevons, one of the founders of economic marginalism, guessed the existence of this paradox by noting that the introduction of ever-more-efficient steam engines brought about an enormous increase in the use of coal.

It is a counterintuitive phenomenon, but still easy to understand in its essence. If I have a light bulb that uses half as much power as another one, I may pay less attention to how long I leave it turned on, because I know it uses less electricity, or I might tend to use the "saved" electricity for something else. This principle works for other forms of energy consumption. For example, if my car gets better gas mileage, I may feel free to drive more. In other words, *a reduction in the unit cost of energy because of greater efficiency may push me to consume more of it.* As a consequence, every energy plan should always assume that some portion of the efficiency gains will always be lost because of this paradox. Ideally, the energy plan should be designed to mitigate this effect as much as possible. The problem is that it is very difficult to evaluate the size of that loss.

The third trap is what is known as the "spillover effect." It consists of the widespread diffusion of new tools made possible by the invention of more energy-efficient devices, or simply by new technologies. In this sense, the spillover effect is partly the other face of the efficiency paradox. For example, the invention of much more efficient devices for air cooling contributed to making air-conditioning not just a luxury item but also a widely popular feature in much new housing. In another field, the technological advancements and cost reductions in personal computers made it possible for them to become a common feature of our houses and offices, thus contributing to substantial increases in the demand for electricity. Because of its unpredictable nature, the spillover effect is even more difficult to calculate than the efficiency paradox.

A fourth trap comprises what can be thought of as "indifference to delayed returns." Many efficient-energy devices are more expensive at the beginning and much cheaper later. However, consumers usually look at the immediate cost to be incurred and are suspicious of or indifferent toward the promise of future savings.

Indeed, in our societies we tend to give greater weight to the initial investment than to diluted savings over time. The only exception is that a significant proportion of consumers seem willing to accept a larger initial disbursement if it is paid back with savings that can be obtained in a very short time.

The problem with this indifference is that it impedes a new energy efficient device in increasing its market penetration and thus reducing its unit cost by economies of scale, a factor that is essential for its widespread diffusion. For this reason, many promising new inventions may remain for years, or even for decades, niche tools for extravagant (or rich) people, without making a dent in the broader market. Yet most efficiency advocates wrongly assume in their plans that consumers will behave according to the advocate's idea of rationality, buying every tool that ensures a payback in a reasonable period of time.

There are many other traps concerning energy efficiency, but I will mention just two more.

The fifth trap is the "policy gap," which is made up of two elements. The first part is the frequent lack of energy knowledge on the part of politicians, which makes them prey to oversimplifications, mistakes, and empty slogans regarding energy policy. The other element is the need for energy regulatory policies that create incentives for energy efficiency and remove existing barriers and disincentives from existing electric utility rates and tariffs. According to former U.S. state and federal energy regulator Branko Terzic, "This means regulatory policies which provide incentives for efficiency in acquisition of fuels and assets, conversion technologies and operations, transmission and distribution, and end use."[13]

The kinds of regulatory policies contemplated here include allowances for earned return by the electric utility on investments in energy efficiency investment, decoupling of utility profits from sales volumes and the introduction of real time pricing to allow for economic pricing signals to reward efficiency investment appropriately.

The lack of informed government policy and regulation has often made it impossible to devise serious, realistic, and holistic plans, which include comprehensive energy efficiency programs.

The final and most dangerous trap (to me) is something that I have repeated several times in this book: cheap energy. When consumers have energy for next to nothing, it is practically impossible to move them toward responsible consumption habits or to build political support for energy efficiency measures.

The Need to Insist on Energy Efficiency

All this said, energy efficiency is a fundamental pillar of any serious energy strategy. As explained by Terzic, "Improved energy efficiency all along the energy conversion and usage path brings the twin benefits of mitigation of greenhouse gas production and the lowering of consumer energy bills in the short run and long run."[14]

The United Nations Development Program's 2000 energy assessment estimated that industrialized countries could achieve cost-effective energy efficiency gains of 25–30 percent by 2020. This is a reasonable target. Even if we discount it by half to be conservative, we still have a net gain in energy through savings that cannot be matched by new production either by nuclear power or any of the renewable sources in the same time frame.

What's more, a stronger focus on energy efficiency by the industrialized countries could bring additional benefits, in that it would make more energy-efficient tools and technologies available to emerging countries to use in their own development. Consequently, it would make the inevitable process of growth in those countries less energy and carbon intensive.

So, what should we do?

Certainly, energy efficiency requires intensive and widespread education for consumers—individuals, families, and especially businesses, where efficiency training can have the greatest effect—about the savings they can achieve by purchasing one device rather than another. In many cases, this process could be promoted by mandating deadlines for the elimination from the market of less energy-efficient devices, as soon as it is proven that there are effective substitutes for them. In other cases, efficiency standards defined by law are a better instrument to impose the required targets over a certain period of time.

There are other means as well. For example, I cannot imagine metropolitan areas of the future still being overrun by vehicles with unnecessarily large engines, low mileage, and high emissions. This is not only a problem of energy consumption, local pollution, and greenhouse gases but also an obvious issue of space, lost time (for instance, looking for a parking place), and many other difficulties tied to such an irrational and absurd habit. So it seems reasonable to me to imagine setting a deadline (let's say, eight years from now—time enough for the automobile industry to adapt and for consumers to change their car-driving habits) after which only small and medium-size cars with high gasoline mileage per gallon and reduced emissions would be allowed to enter downtown areas.

In addition, governments should boost the energy efficiency of industrial plants and grids by allowing companies that are able to produce the same output with less energy to retain a part of the savings, either through fiscal incentives or through a carbon tax imposed on all market operators.

Energy efficiency policies may also rely on subsidies, but according to a very strict and focused framework. It would be a mistake to scatter subsidies, the way that most policies do during periods of mounting emphasis on energy conservation. Subsidies should be tied carefully to the specific efficiency outcomes that new tools are supposed to achieve. In a world where public resources are limited anyway, it is of the highest importance to set priorities for public spending, evaluating carefully which actions will have the best payoff in terms of energy savings and emissions. In other words, we must evaluate costs and benefits carefully. Furthermore, incentives should be reviewed periodically to ensure that we are still rewarding provisions that are in fact more efficient.

Above all, our long-term plan to favor energy efficiency must be decoupled from the future market behavior of fossil fuel prices. We cannot take an interest in efficiency only when the price of energy skyrockets, and then forget it all as soon as prices plummet—as it has always been in history. The world should be aware that energy prices are cyclical, and every boom phase of the cycle tends to impose a higher price level even after the boom has gone by.

We also need to bear in mind that in the world today—with a burgeoning middle class throughout the world and many new foreign consumers starting to imitate the wasteful old ways of Westerners—we will need to pay careful, constant attention to the ways that energy is used.

A last historical lesson needs to be recalled. As Peter Odell has noted, while cheap energy before the 1970s contributed to the "revolution of rising expectations," the eventual jump in oil and energy prices in the 1970s "terminated the perception of energy as a near-costless input to economic and societal developments in most parts of the world."[15]

This had a beneficial effect on our planet. As a matter of fact, as Odell pointed out:

> The continuation post-1973 of the abnormally high growth rate in energy use since the late 1940s would have required an annual global supply of energy of over 22 gigatons of oil equivalent (Gtoe—billions of tons) by 2000. In sharp contrast . . . the actual growth rate of only 1.7 percent a year from the 1973 base led to a use of energy in 2000 of less than 10 Gtoe.[16]

It is true that energy consumption has grown since then in absolute terms. But think what world energy consumption would have been without that effort in terms of higher energy efficiency!

Once again, in the twenty-first century, responsible governments should reproduce the situation that—in the 1970s—terminated the perception of energy as a near-costless input. I will deal with this issue in the Conclusions.

This path may be unpopular, but I think it necessary that world leaders should guide their people to understand that energy is not an eternal free entitlement and that we cannot waste it without consequences for the Earth's future.

Conclusions: How Can We Escape Our Energy Trap?

At the end of this brief journey through the world of energy, it is worth pausing to summarize the salient issues and to draw some conclusions.

Petroleum, Coal, and Natural Gas: The Intractable Reasons for Their Continuing Dominance

As we have seen, the incontestable advantages of fossil fuels (oil, coal, and gas) as sources of energy simply blow away the competition:

- Fossil fuels release much more energy per volume than any other known source except nuclear power.
- They are available when we need them.
- They can be bought and stored for future use.
- They can be transported from one country to another.
- They are flexible enough to provide several forms of secondary and tertiary energy, from heat to electricity to transportation fuel.
- Except for short periods of time, they have always been much cheaper than any other source of energy.

Doomsayers notwithstanding, shortages will not end the golden age of fossil fuels any time soon. The apparent shortage of crude oil in the first decade of this century, for example, was merely the result of meager investments in exploration and development during the 1980s and '90s. As I explained in the chapters on fossil fuels, not only are their known resources still huge but they are also bound to increase over time in step with technological advances in exploration and recovery.

The dire projections of "Peak Oil" theory are nothing more than a reformulation of the traditional Malthusian growth model. Long since discredited as a predictor of population growth, Malthusianism extrapolates the future on the assumption of a fixed and precisely measurable stock of resources. This assumption in turn requires that knowledge and technology be assumed to be fixed and incapable of enlarging the available stock of resources. The only permitted variable is resource consumption, which is assumed to grow exponentially until the fixed stock of resources is all used up. But oil resources are not comprehensively and conclusively measurable. All that we can infer from our current limited knowledge is that they are still plentiful and that their discovery and extraction depend primarily on prices and technology.

At the same time, the most polluting energy source, coal, is anything but in decline, and it will be difficult to curb its consumption patterns, not only because of its enormous availability but also for reasons of simple opportunity. Almost 90 percent of it is used in countries that have vast, low-cost reserves on their own territory, notably the United States, China, and India. They will find it difficult to do without coal.

Finally, the impressive rate of growth in the world energy share of natural gas is expected to continue over time, thanks to its vast reserves and its attractiveness as the fossil fuel that has the least impact on the environment and the highest efficiency in generating electricity. These features will probably vault natural gas into first place among the world's primary energy sources during the twenty-first century.

Given this general picture, how can we realistically challenge the preeminence of fossil fuels, no matter how convinced and alarmed we are that they are wreaking grievous damage on our planet and its atmosphere? How can we hope to succeed when so far no alternative source except nuclear power (and that only for the production of electricity and heat) has proven itself able to make a dent in the oligopoly of fossil fuels?

In other words, how can we free ourselves from our energy trap?

Clean Sources in Crisis: Nuclear and Hydroelectric

To make matters worse, the only complementary sources to fossil fuels capable of making a significant impact on world energy supplies are in a sad state of repair.

In spite of the prospects of a rebirth announced at the beginning of the twenty-first century, the contribution of nuclear power to global electricity generation continues to recede: whereas in 1990 it represented almost 17 percent of world electrical production, by 2008 it had dropped to 15 percent. In reality, the rebirth is more on paper than on the ground.

The nuclear plants under construction worldwide will likely compensate for the many old plants that will be closed in the coming decade. Although many will have their lives extended, others will inevitably be shut down. By 2030, only China, India, and perhaps Russia, South Korea, and Japan are likely to have significant development programs. Too many factors play against nuclear power elsewhere in the world at this time:

- Local communities oppose the construction of new plants.
- We have failed to solve the problem of geological storage of highly radioactive wastes (especially in the industrialized world).
- A very long time will be needed to certify and launch the first fail-safe power plants.
- Upfront costs for building up new nuclear reactors are extremely high and may deter many private investors to bear them, unless governments ensure strong economic, financial, and regulatory support.

These problems are part of the reason for the very long time between the decision to build a new nuclear power station and its coming on line. Power plants built in the last fifteen years on average have required more than sixteen years to become operational (except in Japan and China). As in the past, the budgets on which the work is based have constantly been shown to be optimistic. For example, a power plant under construction in Finland saw its initial budget more than double in a few years, raising again the question of whether nuclear power is truly competitive on a real cost (not hypothetical cost) basis. This does not take into account the fact that soon those countries that turned to nuclear power enthusiastically in the past will find themselves having to pay costs that were not budgeted to decommission the old power plants. Indeed, only a small fraction of these costs was ever accounted for in the initial planning.

With the lack of solid answers to these problems and the doubts about nuclear power, time is not on its side. It risks having its actual rebirth slide past 2030, thus worsening the prospects of winning the long-odds bet that we can achieve sustainable energy in our century.

Hydroelectric energy is not in good shape, either. The construction of large dams is now under attack in many parts of the planet for social and environmental reasons. New projects are taking shape in developing countries, mainly China and Latin America, but not without widespread protests by the local people and by environmental and human rights organizations. At the same time, drought and insufficient rainfall are threatening the productivity of existing installations. Moreover, much of the hydroelectric potential available in the world is not actually exploitable, because it is concentrated in areas that are too far from the centers of consumption (e.g., in Africa), where the two are separated by hundreds of miles of thick forest or deserts.

For these reasons, clean and renewable hydroelectric power continues to see its share of the world electrical power generation shrink, from 20 percent on average in the 1980s to 16 percent in 2008. And its contribution to the world's primary energy is struggling to maintain a meager 2 percent of the total.

The Uncomfortable Truth about Today's Limits on Renewables

A true energy revolution cannot do without sources that can produce great amounts of energy. The last great energy transition in human history took place between the second half of the nineteenth century and the first decades of the twentieth, with the spread of coal and steam turbines, the advent of oil, and later the emergence of natural gas. Human technology triggered this successful transition by exploiting energy sources capable of replacing previous sources—mainly biomass such as wood and vegetable matter—in terms of quality and quantity.

The principal problem concerning renewables is just this: They have a very low energy and power density, so they can produce only limited amounts of energy. In part, this is a technological problem because we still cannot exploit those sources fully, just as we cannot yet accumulate and store electrical energy on a large scale. Perhaps in the future we will be able to do this, but the road ahead is still long and full of difficulties.

The problem of the intrinsic amount of energy deliverable by each source is closely connected to the specific cost of the alternatives to fossil sources. Detractors of clean energy use this argument most often. True, these sources cost more. However, even assuming that society were willing to bear the greater cost in the name of sustainability (itself a reason for a shared course of action), the basic problem remains. They can provide us only modest amounts of energy. They are certainly not in a position to respond to the energy needs of humanity over the next two decades. In this time span, we can only hope that the most competitive alternative sources will take over some part of the growth in the total energy demand, to avoid turning to fossil sources for the entire additional requirement.

Certainly, the boom of renewables—wind power, solar energy, and biofuels in particular—in the early 2000s has aroused many hopeful expectations. But we cannot forget the basic numbers and further problems that characterize those renewable sources.

First, the numbers:

- Wind power still represents only 0.1 percent of total primary energy consumption and about 1 percent of electricity production.
- Solar energy falls even farther behind, accounting for less than one-thousandth of primary energy consumption and much less than 1 percent of electricity production.
- Finally, biofuels for transportation represent a little more than 1 percent of global consumption of transportation fuels.

Many will argue against these numbers by citing very specific cases where renewables play a much bigger role. So, wind power supporters will point to Denmark, where wind energy satisfies one-fifth of the total electricity demand and where wind "works" for more than 4,000 hours per year—twice as much as in other windy areas. Biofuel advocates will point to Brazil, where bioethanol can be produced at prices that cannot be matched by any other country (thanks to unique weather conditions, the availability of cultivatable soil and water, low labor costs, the adaptability of sugarcane to the climate, and thirty years of government incentives).

But the truth is that the world is neither Denmark nor Brazil.

These two countries enjoy unique conditions that cannot be replicated worldwide. It would be the same as saying that Saudi Arabia is typical when talking about oil, the United States when talking about coal, or Russia when addressing natural gas. Nature has placed in those three countries about 25 percent of world reserves of those resources, just as it has supplied Denmark and Brazil with the best conditions in the world for wind energy production and the production of biofuels.

Second, we must acknowledge that most renewables present additional systemic problems that are usually underestimated.

Our electrical systems were planned and built to satisfy the needs of production plants of great size whose performance and operating capacity can be predicted with reasonable certainty. The introduction of relevant shares of wind and solar energy, both of which are characterized by intrinsic variability of supply, would completely change the rules of the game. To begin with, the balance between supply and demand of electric power would become more complex, riskier, and more expensive. In fact, it would require significant development of natural gas backup storage capacity as a substitute for wind and sun when they are not available. Also, it would require a different voltage for transportation and transmission grids than the one that prevailed in the twentieth century. Then, wind and sun, like hydroelectric power, also suffer from a "geographic misfit." They are available mainly in areas far away from the large centers of consumption (think of the Sahara Desert), and this makes the construction of electric lines even more demanding (in both economic and engineering terms). In sum, a greater share of wind and solar power would require massive investments to rebuild our electric systems.

And this would still not be enough. Management systems and networks not only would need to be further developed but also would have to become more "intelligent"; the consumers of electricity themselves would have to adapt to an output that would no longer be predictable and standardized.

Most of these problems probably could be resolved during the twenty-first century, but not in the next twenty years.

Turning to biofuels, their prospects, too, are less rosy than most of their advocates claim.

With traditional crops, enormous areas are required to get modest amounts of biofuel, because once again the energy and power density of any vegetable fuel is extremely modest. Furthermore, the massive use of fertilizers, water, and even energy for the large-scale development of biofuels creates a sustainability problem, making their overall environmental, energy, and emissions balance ambiguous or even negative. Finally, competition between biofuels and food for the use of land and crops (e.g., in the case of cereals) could have a potentially devastating impact on the cost of many foodstuffs, worsening the already precarious conditions of the poorest populations in the world. The first warning signs have already arisen, and we are only at the beginning.

As a consequence, with the exception of Brazil's bioethanol, first-generation biofuels appear to be a remedy that is worse than the problem they are supposed to solve. Meanwhile, second-generation biofuels are still on the drawing board because of the need to solve their problems of cost, energy balance, and logistics.

For all these reasons, the uncomfortable truth about today's renewables is that they cannot free us from our energy trap over the next few decades. The time will come when they could become significant resources, but as complementary sources rather than alternative ones. The reality is that they will not play even a complementary role until well after 2030—and then only if we act now to improve energy efficiency and to fund research and development.

What to Do?

The big uncomfortable truth of this picture is that the large-scale push under way throughout the world to develop clean and renewable energy runs the risk of doing little more than compensating for the problems of nuclear power and hydroelectric (the two sources with zero carbon dioxide emissions), without making the smallest dent in the use of fossil sources. On the contrary, the use of fossil fuels may increase even more, at least in the next few decades of this century. We seem doomed to lose the battle to ease the heavy carbon and pollution burden on our planet.

As I said at the beginning of this book, I belong to the school of thought that disagrees with the post-Malthusian view of the world. Yet, we will have to face the unprecedented explosion of people who will demand the same (or similar) standards of living that only

Westerners enjoyed until recently. This will pose a threat that requires a profound change in our relationship with energy.

World leaders must make serious choices with gravitas, a rare commodity in the public arena, where debate is too often fed by slogans, irresponsible simplifications, and rapacious inventors and investors who easily turn into "green swindlers." A new energy and environmental policy cannot be the result of media gossip or attractive but unfounded hypotheses. Instead, it must start realistically, setting an ambitious course based on at least seven fundamental points. Let me try to summarize them:

1. We will have to live with fossil fuels for a long time to come. Therefore, we need an immense effort in research and innovation and in regulation and public awareness, to minimize polluting potential of fossil fuels at the source and to reduce their threat to the climate danger, from extraction through combustion.
2. At the same time, we cannot give up on those relatively clean sources—nuclear and hydropower—that may help curtail the expansion of fossil fuels. We may not like them, but the only choice we have is between something we do not like and something that is worse (an even greater share of fossil fuels).
3. We need incisive action to increase efficiency in the conversion and final consumption of energy. This will come not only from collective voluntary effort but above all from careful regulatory choices.
4. We need a gigantic effort in scientific research to focus on innovative technologies that can make the renewable sources real alternatives, that is, competitive in terms of energy density and power density, as well as cost over the medium and long terms.
5. A combination of binding regulations and focused public subsidies will be needed to realize the points described above. Public resources must not be wasted on scattershot interventions. Instead, they must be used with focused logic and a set of priorities aimed at sustaining serious and potentially innovative research. They must support only those industrial applications that reduce pollution and emissions the most for the same amount of energy produced.

6. A serious energy and environmental policy is a long-term effort. It must step back from the hysteria caused by high prices for oil and gas, which vanishes as soon as prices return to moderate levels, along with interest in alternative energies. Launching and sustaining a true energy transition cannot be done in short-term emotional bursts.
7. Finally, at least among the principal consumer nations, international coordination of energy and environmental policy is important to avoid having the results obtained in some countries with great effort become a disincentive for others, compromising the final result (the classic problem of the free-rider).

Unfortunately, it is relatively easy to put on paper some basic principles for a sound energy policy, but highly difficult to transform them into a sound and effective long-term energy plan.

How Do We Do It?

In my career, I have had the opportunity to study dozens of serious energy plans, based on sound principles and extensive cost-benefit analyses. Of course, none of them was perfect, but among the many other silly plans out there, these were serious, complex, and well-articulated and, at least in general, their targets were achievable. Yet none of them has really succeeded in having a significant impact. Why did this happen?

My final conclusion is that it is nearly impossible to devise a comprehensive energy plan while not being in control of the single, most upsetting variable of the energy world: the price of fossil fuels—of oil, in particular.

We cannot fight against something that for most of its history has been so affordable as to displace any other option and that may continue to be so in the future. As we have seen, the most legitimate proposals for energy efficiency, as well as most attempts to develop affordable renewables, have been killed by unanticipated crashes of the prices for oil and other fossil fuels. Without solving this problem, I fear that any plan concerning energy, no matter how detailed and well devised, will be bound to fail.

So, instead of proposing an additional plan among many others, I prefer to conclude my book by stressing that the prerequisite for any plan to have a chance of success is *to find a way to terminate the perception of energy as a near-costless input.*

In our times, there is a widespread view that our planet will resolve this problem by itself. In fact, most are convinced that the natural tendency is for energy demand to grow without end. Because oil and other fossil fuels (at different stages) will eventually run up against the limits of a finite resource, fossil fuel prices will inevitably increase, obliging mankind to find other sources of energy. I would not bet on this scenario, however, for the reasons I have explained in this book. Prices will continue to run up and down in the future, and probably their average level will move higher than in past decades. Yet, in bullish cycles, they will still fall low enough to derail any serious attempt to boost energy efficiency and produce energy alternatives.

In my view, *the only way to solve this problem is a concerted effort to artificially increase fossil fuel prices,* through a combination of two main tools.

First, as I alluded in chapter 12, I am in favor of an immediate, moderate carbon tax, running in parallel with a cap-and-trade system.

Most economists may object that an immediate cap-and-trade program would be much more effective and less distorting in the long term. I totally agree with them. But if we want to get results *as quickly as possible,* a carbon tax would prove more effective.

In my experience, industrial companies tend to focus mainly on short-term results and dangers, while they pay little attention to long-term threats—in spite of lip service to the contrary. Consequently, company leaders react strongly to short-term challenges and let their successors take care of the future problems. That is why, for example, the world's first commercial project of carbon capture and storage came online in Norway (at Sleipner; see chapter 4) in 1996: Norway had imposed a carbon tax on offshore hydrocarbon extraction in 1991, which obliged oil companies to search for ways to minimize the effect of this burden.

Furthermore, what makes cap-and-trade systems much slower in deploying their effects is politics. Usually, when they are first introduced they are all but perfect systems, because they are the results of both too much uncertainty about the way to allocate quotas among emitters and too much lobbying by vested interests to receive free emission allowances. Consequently, the first years of a cap-and-trade regime may turn out to be useless with regard to the goal they aim to achieve, as the European experience has shown.

In my opinion, a carbon tax represents an impetus that requires an immediate response by industry, while cap-and-trade systems do not. A cap-and-trade system becomes most effective after many years of trial and error, at the end of which it may ensure the perfect functioning of a market for carbon emissions—a system of emission allowances correctly tailored to the characteristics of a specific country. That is why I argue that the best choice for a government would be an immediate carbon-tax associated with a cap-and-trade system. Only when the latter succeeds in deploying its overall effects, may the carbon tax be cancelled. *In sum, I prefer to escape the dictatorship of "or-or," and to accept the wider options opened by the democracy of "and-and," with a system where carbon taxes and cap-and-trades live together for as long as it is necessary.*

However, there is a minefield of problems in adopting carbon regulations. If only one country or group of countries decides to impose these regulations, whether on the industries transforming fossil resources into usable products or on the products themselves (through strict quality standards), there is a significant risk of driving production to other countries that have weaker standards and of increasing the cost of many goods (goods that depend on energy) in the domestic economy. This, of course, would benefit those nations using less expensive, presumably more polluting, energy systems. Above all, the global environmental balance would suffer, through the relocation of the polluting manufacturing industries and their emissions to countries that do not consider these problems a priority, thus undoing the effort undertaken by virtuous countries. This is a serious problem, which must not be trivialized.

It is difficult for us Westerners—and for Americans in particular—to ask emerging countries such as China and India to limit their ambitious plans for growth for the sake of avoiding disastrous climate effects. Many other countries of the world have a right to complain that we have guzzled resources for many decades and are now asking the world to go on a diet just as they are sitting down at the table. They can point out that they are doing better than the West did when we started, bringing up the forgotten truths of the pollution and emissions, as well as human lives and unsustainable working conditions, during the early Industrial Revolution.

A sacrifice by the West, then, is required. As a first step, all Western countries could self-impose carbon taxes and cap-and-trade systems on their activities. This would be proof of good faith for the rest of the world. After a certain period of time, the developing countries should apply the same systems, according to an index of development that may determine a different schedule for different countries.

Probably, the fairest system to distribute emissions cuts would consist of looking not at a country-level, but at high individual emitters wherever they live. As explained by an important and innovative study by a team of experts, high individual emitters are those whose emissions exceed a universal individual emission cap. Looking at the problem this way, it is possible to identify groups of high emitters in four main areas of the world (the United States, the OECD minus the United States, China, and the non-OECD minus China) that by 2030 will represent the parties most responsible for the 43 billion tonnes of carbon dioxide emitted from fossil fuels in the "business as usual" scenario of the IPCC. To lower that level of emissions and stabilize it around 30 billion tonnes, it would be necessary to cap individual emissions to 10.8 tonnes of CO_2 per year—a ceiling that would engage around 1.13 billion people out of more than 8 billion globally at that date. Once the principal high emitters are identified, each country should be free to enforce the policies it may deem more appropriate to curb its own emissions.

In any case, if the concerned developing countries do not move in the same direction first adopted by industrialized countries after the "period of grace," something else needs to be tried.[1] And perhaps the only way to resolve this problem, painful as it might be, would be to impose a carbon tariff, for example, customs fees on goods coming from countries that do not respect the same environmental standards, in order to avoid having the responsible attitude of one country or part of the world simply translate into an advantage for its competitors. It is easy to understand that this is not a simple path.

For years, we have been discussing applying such provisions against nations that use child labor or do not provide for occupational health and safety of their workers, so far without result. But the health of the Earth calls for something new.

The third tool I favor to increase fossil fuel prices is the introduction of a "mobile excise tax" on end-use products derived from fossil fuels. Such a tax should be applied when prices fall below the level that sustains energy efficiency, technological innovations, and research programs into alternative sources of energy. Considering the concurrent burden posed by the aforementioned carbon tax, I estimate that under current conditions the level below which the excise tax should be activated would be around $60 per barrel of oil, or the equivalent value (in terms of energy parity) for coal, while a discount should be applied for natural gas due to its more environmentally friendly nature. Of course, conditions may change over time, and this would require a constant adaptation of the tax to changing circumstances.

A Desirable Future

I know that my proposals for two taxes and a cap-and-trade system on fossil fuels will raise much skepticism. But I am convinced that introducing them will not harm economic growth. On the contrary, if they are properly designed, they will trigger the ingenuity of real entrepreneurs and managers, which in turn will accelerate innovation and change. As I have often repeated in this book, cheap energy and energy innovation cannot coexist. And unfortunately, the current state of knowledge and technology do not offer us any shortcuts to escape our energy trap.

Yet, compared to the past, many elements warrant some optimism and should push us to "seize the moment."

First, research and development into energy efficiency, renewables, as well as environmental remediation and emissions abatement, were once subject to Darwinian cycles of economic competition. Thus, when the price of oil and other fossil fuels went through the roof, private and public investment poured into research, only to evaporate when prices plunged. Such a cycle of higher and lower costs could repeat itself in the coming years, but now environmental and climate preservation awareness has reached much of our planet. This should keep the laboratories of the world from finding themselves devoid of resources to continue their research, even if oil prices fall. In any case, to avoid the worst effects of such a fall, I reiterate my belief in the three pillars of a carbon tax, cap-and-trade, and mobile excise tax.

Second, available public resources are being aimed at research and development into a larger variety of alternatives to fossil fuels than ever before. Over the last fifty years, nuclear power absorbed almost 90 percent of government research financing, leaving only crumbs to those research threads that deserved broader consideration, but this is no longer the case.

Third, the growing interdisciplinary blending of biochemistry, quantum physics, molecular engineering, and many other fields taking place in the field of energy could generate solutions that cannot be foreseen today. This is a relatively new development, heralding many possible evolutions. We should encourage and sustain interaction among many different sectors of scientific research.

Fourth, the Western oil industry has a great interest in massively investing in new forms of energy, because its traditional role is in decline. Over an uneven but one-way path in the last fifty years, producer countries have reappropriated the reserves and many of the contractual advantages that at one time made the major Western oil companies absolute lords of oil and gas worldwide. What is left to the oil companies today is a minimal part of those reserves, and even that is at risk.

These companies cannot face their future without exploring new forms of energy, but they must also look for advanced technologies to mitigate the environmental and carbon footprint of their current core businesses (oil and gas), without which their role could even face a more rapid decline in many parts of the world.

Finally, even several big oil-producing countries such as Saudi Arabia and the United Arab Emirates are investing in renewable sources of energy, both to find alternatives for the production of their own electricity, thus freeing more of their precious oil and gas resources for export, and to ensure that their future generations can continue to rely on an abundant oil and gas endowment. By the same token, some big emerging oil guzzlers—above all China—are also investing massively in renewables, as well as in technologies that may help reduce the environmental and carbon effects produced by their startling economic growth.

These elements form the leading edge of a long wave, which needs to be supported through the measures that I have indicated. I reiterate the notion: Only if the world succeeds in banishing the notion of energy as nearly free will that wave of innovation water fertile ground from which a major energy revolution could grow.

We have never seen anything like this in the past, nor do we know when one or more of the many seeds sown will germinate vigorously. What we can reasonably expect is that if the sowing continues, sooner or later the seeds will germinate, probably when we least expect them. Then we will finally be able to rely on "recipes for better cooking" that will free us from our energy trap.

Notes

Introduction

1. Robert A. Muller, *Physics for Future Presidents: The Science Behind the Headlines* (New York: W. W. Norton & Company, 2008), 67.

2. On this issue, see: J. R. McNeill, *Something New Under the Sun: An Environmental History of the Twentieth-Century World* (New York: Norton, 2000).

3. Paul Romer, "Economic Growth," in *The Concise Encyclopedia of Economics*, edited by David R. Henderson (Indianapolis, IN: Liberty Fund, 2007). http://www.econlib.org/library/Enc/EconomicGrowth.html.

4. Thomas L. Friedman, *Hot, Flat, and Crowded: Why We Need a Green Revolution—And How It Can Renew America* (New York: Farrar, Straus and Giroux, 2008), 55.

5. Moses Naim, "Can the World Afford a Middle Class?" *Foreign Policy* 165 (March–April 2008): 95–96.

Chapter 1

1. Daniel Yergin, *The Prize: The Epic Quest for Oil, Money and Power* (New York: Simon & Schuster, 1991), 194.

2. Leonardo Maugeri, *The Age of Oil: The Mythology, History, and Future of the World's Most Controversial Resource* (Westport, CT: Praeger, 2006), 130.

3. For more information on the theories of Hubbert and his followers, see Maugeri, *Age of Oil*, 201–6.

4. One barrel contains about 42 U.S. gallons (159 liters) of petroleum. The barrel became the worldwide unit of measurement for petroleum at the dawn of the modern age of oil. The first drillers in Pennsylvania (1859–1860) collected petroleum in 42-gallon wooden barrels, which were used at the time for aging whiskey. Although crude oil is no longer marketed physically in barrels, the unit of measurement remains.

5. U.S. Geological Survey, *World Petroleum Assessment 2000*, http://pubs.usgs.gov/dds/dds-060.

6. Data calculated by the author based on the IHS database, https://peps.ihsenergy.com (by subscription only). See also the IHS website, www.ihs.com/energy.

7. Historically, the recovery rate has been very low. For example, in 1980 it hovered around 20 percent.

8. Morris Adelman, *The Genie out of the Bottle* (Cambridge, MA: MIT Press, 1995), 15.

9. Leonardo Maugeri, "Squeezing More Oil from the Ground," *Scientific American* (October 2009), 56–63.

10. The Seven Sisters were Standard Oil of New Jersey (which became Exxon and is now ExxonMobil), Shell, the Anglo-Iranian Oil Company (BP), Standard Oil of California (Chevron), Standard Oil of New York (Mobil, which merged with Exxon in 1999), the Texas Oil Company (merged into Chevron in 2000), and Gulf Oil Corporation (acquired by Chevron in 1983). Between the 1950s and the 1970s, the seven companies controlled more than 50 percent of the world reserves of crude oil and more than 70 percent of the refining and distribution capacity worldwide.

11. Maugeri, *Age of Oil*, 77–91.

12. OPEC was established in Baghdad in 1960. Today twelve countries belong to it: Algeria, Angola, Ecuador, Iran, Iraq, Kuwait, Libya, Nigeria, Qatar, Saudi Arabia, the United Arab Emirates, and Venezuela.

13. Maugeri, *Age of Oil*, 6–11.

14. David Hobbs, "New Institutions and International Dialogue," *Oil* 7 (October 2009), 22–25.

15. Maugeri, *Age of Oil*, 103–4.

16. Maugeri, *Age of Oil*, 112–14.

17. Adelman, *Genie out of the Bottle.*

18. Maugeri, *Age of Oil*, 188, 206.

19. See Robert Mabro, "The International Oil Price Regime: Origins, Rationale, and Assessment," *Journal of Energy Literature* 11, no. 1 (June 2005), 3–20.

20. Leonardo Maugeri, "Understanding Oil Price Behaviour through an Analysis of a Crisis," *Review of Environmental Economics and Policy* (Summer 2009), 1–20.

21. Cambridge Energy Research Associates (CERA), *"Recession Shock": The Impact of the Economic and Financial Crisis on the Oil Market* (Cambridge, MA: CERA, 2008).

22. Hugh Thomas, *The Suez Affair* (Harmondsworth, Middlesex, England: Penguin Books, 1970), 32.

23. Morris Adelman, "The Real Oil Problem," *Regulation* (Spring 2004): 19.

24. "Cheap Oil: The Next Shock?" *Economist*, March 4, 1999, 21–23. Available at http://www.economist.com/world/displaystory.cfm?story_id=188181.

25. Brian O'Keefe, "Here Comes $500 Oil," *Fortune*, September 22, 2008. Available at http://money.cnn.com/2008/09/15/news/economy/500dollaroil_okeefe.fortune/index.htm.

26. Leonardo Maugeri, "Two Cheers for Expensive Oil," *Foreign Affairs* 85, no. 2 (March–April 2006), 149–61.

27. CERA, "*Recession Shock*."

28. Ed Morse, "Low and Behold," *Foreign Affairs* 88, no. 5 (September–October 2009), 36–52.

29. On the evolution of petroleum pricing systems, in addition to Mabro, "International Oil Price Regime," see Bassam Fattouh, "The Origins and Evolution of the Current International Oil Pricing System: A Critical Assessment," in *Oil in the 21st Century*, ed. Robert Mabro (Vienna: OPEC, 2005), 41–100.

30. Annualized average of daily data published in U.S. Commodity Futures Trading Commission, *Commitments of Traders: Historical Reports* (Washington: GPO, 2008). Available at http://www.cftc.gov.

Chapter 2

1. For a comprehensive history of coal, see Barbara Freese, *Coal: A Human History* (Cambridge, MA: Perseus, 2003).

2. By convention, the three fossil fuel sources are measured thus:

- 1 metric ton (tonne) of petroleum = 10 million kilocalories (kcal)
- 1 tonne of coal = 6 million kcal
- 1 tonne of liquefied natural gas = 12.3 million kcal

3. Coal contains up to 72 of the 116 chemical elements on the periodic table in various mineral compounds. Used for the first time in 1868 by the Russian Dmitry Ivanovič Mendeleev, the periodic table of elements is a chart in which the various chemical elements are classified by the configuration of their electrons so as to observe the periodicity of their characteristics. Chemical elements are homogenous substances that cannot be divided chemically into smaller units, because they consist of atoms of the same type.

4. The most used parameters are the degree of fossilization, the chemical composition, the content of ash and sulfur, and the intended use. The ash and sulfur contents are very important values, because their presence beyond certain limits negatively affects the uses of coal and its price.

5. By comparison, petroleum has an average thermal value of 10,000 kcal/kg.

6. *Coking coal* possesses the chemical-physical qualities ideal for transformation into *coke*. Coke is a solid product obtained from the dry distillation

of fossil coal (*coking* is the refining process). Coke can also be obtained by condensing heavy petroleum residues. Coke features low levels of humidity, ash, and volatile materials. It is used in blast furnaces to produce steel. On the other hand, *steam coal* includes all bituminous coals not classified as coking coal, which are used mainly to make steam, especially in the thermoelectric sector.

7. Comparing physical tons of coal to *ton-equivalents of petroleum* (TEP: the amount of energy released by the combustion of one tonne of petroleum), the percentage of world production of coal that is exported rises to about 18 percent. The change is explained by the fact that most exported coal is bituminous, with a higher heat content (higher-value coal); therefore, in going from tons to TEP, there is an apparent weight gain compared to total production (which includes the low-value coal that is not exported).

8. *Coal hydrogenation* involves producing a reaction by mixing coal and hydrogen under high pressure and temperature. The process breaks down the molecules of both reagents and combines them, thus obtaining liquid organic compounds. German chemist Friedrich Bergius was the first scientist to demonstrate the feasibility of this transformation, which he patented in 1913.

9. South Africa was forced to resort to liquefied coal because of international sanctions during the apartheid era.

10. The use of coal in the thermoelectric sector accounts for 63 percent of primary consumption of this fuel worldwide. Coal is also used in the metallurgical sector (iron and steel production), which accounts for about 21 percent, and in chemical manufacturing, 1.3 percent. The remaining 15 percent is divided between the production and sale of process heat or district heating (3 percent); internal consumption and other energy sector transformations (3.3 percent); other industrial sectors (4.1 percent); civil, commercial, and agricultural consumption (3.6 percent); transportation (0.15 percent); and nonenergy uses (0.6 percent).

11. British Petroleum (BP), *Statistical Review of World Energy 2007* (London: BP, 2007).

12. Unless otherwise specified, "tonnes of coal" refers to physical amounts. Converting from physical values to energy values (TEP) is especially complex for coal, because each type has its own heat value and therefore its own conversion coefficient. In general, because a ton of coal contains many fewer calories than a ton of petroleum, converting from a physical measurement (ton of coal) to an energy measurement (ton-equivalent of petroleum) causes a significant reduction in the number of "tons" (as much as 50 percent); this could mislead an inexperienced reader. Therefore, it is always necessary to pay attention to the unit of measure being used (simple tons or TEP) in comparing and evaluating the numbers shown in the text.

13. William Stanley Jevons, *The Coal Question: An Inquiry Concerning the Progress of the Nation and the Probable Exhaustion of Our Coal Mines* (London: Macmillan, 1865).

14. "Developed" and "industrialized" refers to countries belonging to the Organization for Economic Cooperation and Development.

15. Jeff Goodell, *Big Coal: The Dirty Secret behind America's Energy Future* (Boston: Houghton Mifflin, 2006), xx.

16. See the website of the World Coal Institute at www.worldcoal.org.

17. According to a study coordinated by the United Nations, human activity has tripled mercury levels in the atmosphere, bringing the total load to 5,200 tonnes from a preindustrial level of 1,800 tonnes. Meanwhile the rate of mercury deposition on the soil and absorption in the water is between 1.5 and 3 times higher than it was during the preindustrial age. Deposition in the soil went from 800 tonnes to 2,200 tonnes; in water, from 600 tonnes to 2,000 tonnes. The figures are 10 times higher in urban areas near industrial sites. See United Nations Environment Program (UNEP), *Global Mercury Assessment* (New York: UNEP, 2002).

18. If we take a volume of coal and a volume of gas that contain the same amount of energy, burning the coal emits 1.7 times more carbon dioxide than burning the gas. For example, burning the one TEP of coal emits on average 3.9 tonnes of CO_2, while burning one TEP of gas emits 2.3 tonnes.

19. In this case, the thermoelectric sector includes operators who produce both heat and electricity and those who produce heat only.

20. The data are taken from the MIT study cited above.

21. The coal is ground into a powder, most of the particles of which are less than 0.075 millimeters in size.

22. See Stu Dalton, "Advanced Clean Coal," presentation for the California Energy Commission 2007 Workshop, Integrated Energy Policy Report Committee, Sacramento, May 29, 2007.

23. MIT, *Retrofitting of Coal-Fired Power Plants for CO_2 Emissions Reduction* (MIT Energy Initiative Symposium, March 23, 2009), 16.

24. Ken Garber, "Why Making Coal Cleaner Will Take Years," *U.S. News & World Report*, April 1, 2009, 68.

25. For more detail, see Massachusetts Institute of Technology, *Future of Coal*, chap. 5.

Chapter 3

1. The precise term for these gases obtained at the wellhead is *natural gas liquids* (NGL). *Liquefied petroleum gas* (LPG) indicates a mix of propane, propylene, butane, and butene obtained by refining petroleum. The definitions confuse even the experts.

2. See in particular Thomas Gold, *The Origin of Methane (and Oil) in the Crust of the Earth* (Washington: U.S. Geological Survey, 1993). For a general illustration of the issue, see Vaclav Smil, *Energy at the Crossroads: Global Perspectives and Uncertainties* (Cambridge, MA: MIT Press, 2003), 222–23.

3. The U.S. Geological Survey estimates that the "last remaining resources of gas," a total that includes still undiscovered reserves, amount to more than 300,000 billion cubic meters. International Energy Agency (IEA), *World Energy Outlook 2006* (Paris: IEA, 2006), 114.

4. The investment costs for the LNG chain have dropped significantly over the last thirty years, thanks to technical improvements, especially in the liquefaction phase. On the other hand, over the last three years, a general increase in the cost of engineering services and materials has brought about an increase in investment costs. The International Energy Agency estimates that a greenfield project would require investments on the order of $3–$5 billion for liquefaction, $2 billion for transportation, and $800,000–$1 million for regasification. (A *greenfield project* is an industrial installation in an area without other industrial installations; the opposite is called a *brownfield project*.) IEA, *World Energy Outlook 2006*, 51.

5. For the same amount of gas emitted, venting is especially damaging to the environment, because methane has a greenhouse gas effect of about twenty times that of carbon dioxide.

6. Christopher J. Castaneda, "History of Natural Gas," *Encyclopedia of Energy* (Oxford, England: Elsevier, 2004), 4: 230 ff.

7. D. Yergin and M. Stoppard, "The Next Prize," *Foreign Affairs* 82, no. 6 (November–December 2003): 103–14.

8. In 1954, a decision by the U.S. Supreme Court (the *Philips* decision) gave the Federal Power Commission (FPC, later the Federal Energy Regulatory Commission or FERC) the authority to regulate the wellhead price of natural gas going into interstate commerce. The *wellhead price* is the price at the point where the gas is extracted, without transportation and distribution costs. After 1954, natural gas prices were based on the cost incurred by the producer, known as *cost-of-service ratemaking*. This system proved difficult for the FPC to manage, because it required collecting and analyzing producers' costs case by case—too expensive a mechanism from a bureaucratic and organizational point of view. In 1960, the FPC set a single tariff by production area, which often did not represent the real costs incurred by many producers. In 1974, the FPC adopted a single national tariff, given the difficulties of managing a system by geographic areas. Price controls on gas between 1954 and 1978 obviously distorted the development of the American market. The price at which natural gas was sold by the producers was lower than its real market value, especially after the first oil shock drove up the prices of petroleum products competing with natural gas for many end uses. On one hand, this situation discouraged producers from investing in research and development of gas reserves; on the other hand, it created an explosion of domestic demand, setting up an unbalanced situation, which culminated in the winter of 1976/77 with a series of supply interruptions. For more on the history of gas market regulation in the United States, see Malcolm W. H. Peebles, *Evolution of the Gas Industry*

(London: Macmillan, 1980), 59–63, 75–83, or www.naturalgas.org. A very good summary is also contained in Roy L. Nersesian, *Energy for the 21st Century: A Comprehensive Guide to Conventional and Alternative Sources* (Armonk, NY: M. E. Sharpe, 2007), 234–42.

9. For more on the birth and development of the gas industry in the Soviet Union, see David G. Victor, Amy M. Jaffe, and Mark H. Hayes, eds., *Natural Gas and Geopolitics* (Cambridge: Cambridge University Press, 2006), 126–37.

10. The first load of LNG from Lake Charles was delivered to the terminal at Canvey Island, England, in February 1959. See Peebles, *Evolution of the Gas Industry*, 185–94.

11. Eni, *World Oil & Gas Review 2007* (Rome: Eni, 2007).

12. For example, between 1998 and 2005, the United States installed about 170 GW of new generating capacity, more than 95 percent of the new capacity fed by fossil fuels. Europe built almost 60 GW, about 75 percent of new thermoelectric capacity.

13. Combined-cycle gas turbine (CCGT) plants consist of two thermodynamic cycles for transforming heat energy into electrical energy. The first cycle consists of a gas turbine, with exhaust gases at a very high temperature. These gases are sent to a *recovery boiler*, where they provide the heat for a steam cycle.

For the same amount of electrical energy generated by a coal-fired steam plant, a CCGT plant produces one-half the carbon dioxide and nitrous oxides and one-twelfth the amount of carbon monoxide and dust; compared to an oil fired steam plant, about two-thirds the carbon dioxide, half as much nitrous oxides, one-fifth as much carbon monoxide, and one-tenth as much dust.

14. Unlike coal and fuel oil, burning natural gas does not generate significant amounts of sulfur oxides.

15. Ed Morse, "Low and Behold," *Foreign Affairs* 88, no. 5 (September–October 2009), 42.

16. Morse, ibid., 42.

17. The PGC consists of a group of academics and industry experts supported by the Colorado School of Mines. It delivers a report on American gas every two years. See Potential Gas Committee, *Potential Gas Supply of Natural Gas in the United States* (Golden: Colorado School of Mines, 2009).

18. On this issue see: Cambridge Energy Research Associates (CERA), *Gas from Shale: Potential Outside North America* (Cambridge, MA: CERA, 2009).

19. In 1996, the LNG producers were, in the order of exported volume: Indonesia, Algeria, Malaysia, Australia, Brunei, the United Arab Emirates, the United States, and Libya. Today, these have been joined by Qatar, Nigeria, Trinidad and Tobago, Egypt, Oman, and Equatorial Guinea. By 2020, the following could be exporting LNG as well: Iran, Russia, Yemen, Angola,

Venezuela, Norway, Peru, Brazil, Bolivia, Mauritania, Papua New Guinea, and Myanmar; however, it is probable that several of these countries will delay their LNG projects.

20. On this issue see: Gawdat Bahgat, "Prospects for A Gas OPEC," *Middle East Economic Survey,* 52:2 (12 January 2009), 26–30.

21. It is estimated that in 2010–2012, lacking other alternatives, Iran alone will need to reinject 100–200 bcm per year. See Energy Intelligence Group, "Iranian Gas Games," *World Gas Intelligence*, January 31, 2007.

22. Eni, *World Oil & Gas Review 2007* (Rome: Eni, 2007).

23. See in particular: Jonathan Stern, *Continental European Long-Term Gas Contracts: Is transition away from oil product-linked pricing inevitable and imminent?* (Oxford: Oxford Institute for Energy Studies, September 2009, NG 34), 1–24.

24. "Dark Days Ahead," *The Economist* (August 8, 2009), 27–29.

25. Cambridge Energy Research Associates (CERA), *European Gas: Time to Reappraise* (Cambridge, MA: CERA, 2009), 6.

26. See statistics produced by the International Association for Natural Gas Vehicles at www.iangv.org.

27. See Sam Fletcher, "Natural Gas Vehicles Gain in Global Markets," *Oil & Gas Journal* (February 16, 2009), 20.

28. CNG is made by compressing natural gas to less than 1 percent of its volume at standard atmospheric pressure. It is stored and distributed in hard containers (usually cylindrical or spherical), at a pressure of 200–220 bar (2,900–3,200 psi).

29. See www.naturalgasvehicles.com.

Chapter 4

1. Intergovernmental Panel on Climate Change (IPCC), Working Group III, *Climate Change 2007: Mitigation* (Cambridge: Cambridge University Press, 2007).

2. Michael R. Hamilton, Howard J. Herzog, John E. Parsons, "Cost and U.S. public policy for new coal power plants with carbon capture and sequestration," *Energy Procedia* 1, 2009 (www.elsevier.com/locate/procedia), 4487–4494.

3. International Energy Agency (IEA), *Energy Technology Perspectives 2008* (Paris: IEA, 2008).

Chapter 5

1. The *Economist* dedicated a cover story to the "Nuclear Power's New Age" (September 8, 2007).

2. The Soviet Union can claim the first commercial nuclear reactor, in Obninsk, 60 miles (100 kilometers) southwest of Moscow. It came online

in June 1954. The first Western commercial reactor went into operation in December 1956 at Calder Hall in Great Britain, while the first American civilian installation became operational at Shippington, Pennsylvania, in December 1957.

3. The Three Mile Island accident was the only one in the history of United States in which radioactivity escaped. Yet, there were no deaths and no measurable dose reported among the public. Nonetheless, it created a crisis of confidence in the safety of nuclear power in the United States.

4. For up-to-date information about existing nuclear plants, visit the World Nuclear Association website at www.world-nuclear.org.

5. According to the reference scenario in the International Energy Agency's (IEA) *World Energy Outlook 2006* (Paris: IEA, 2006), the installed capacity of nuclear power plants by 2030 will reach 416 GW, but the share of nuclear power compared to other primary energy sources will diminish. An alternative IEA scenario shows greater use of nuclear power helping to contain CO_2 emissions and investments and the sector pushing its capacity to 519 GW by 2030. China and India are expected to increase the share of installed nuclear power capacity by 6 percent and 9 percent, respectively, for a total of 75 GW (compared to the current 9 GW). An overall installed capacity of 93 GW is expected by 2030 in all emerging countries. This would be about one-quarter of the present electrical capacity in the world.

6. There were some thorium (Th) nuclear reactors in India. This fuel was chosen because the Asian country has many thorium mines. However, the high cost of production and the difficulty of treating the irradiated fuel made it so that currently in India there is only one experimental thorium reactor, while the rest of the nuclear power plants are fueled by uranium.

7. An *isotope* has the same number of protons and the same chemical properties as the element that defines it (for example, uranium), but it has a different number of neutrons. This gives it a different *atomic weight*, which is the sum of its protons and neutrons. The atomic weight is indicated by a number next to the name or symbol of the element, for example U-235 or C-14.

8. They are called *thermal reactors*, because the fission reaction of the U-235 is produced by low-energy neutrons, appropriately called *thermal neutrons*. A *moderator* is used to slow neutrons. In reactors with high-speed neutrons, there is no moderator, and the nuclear reactions take place between unslowed neutrons (*fast neutrons*) and the fissionable nuclei. These reactors are still experimental, even though on a large scale. One example is the Superphénix experimental high-speed nuclear reactor, which stopped producing electrical power commercially in France in 1996. Actually, U238 is fissionable by fast neutrons. Also during a normal reactor cycle, some U238 is converted to Pu239 through neutron capture and subsequent decay; this Pu239 is also fissioned by thermal neutrons and in fact provides a significant part of the energy produced in a uranium-fueled reactor.

9. Enrichment is needed only for reactors moderated with light water. If deuterium or graphite is used for moderating the neutrons, natural uranium can be used. The Canada Deuterium Uranium (CANDU) reactors, for example, use deuterium.

10. There are also reactors capable of using natural uranium, for example, the CANDU system. These are heavy-water pressurized reactors developed by Atomic Energy of Canada, Ltd. (AECL) and are already commercially proven. About thirty of these power plants are in operation.

11. The bomb dropped on Hiroshima in 1945 used about 60 kilograms (130 pounds) of uranium-235. Its power was 13 kilotons, equal to 13,000 tons of TNT. The plutonium bomb used on Nagasaki was more powerful (25 kilotons). In general, the core of an atomic bomb can be built with a few dozen kilograms of enriched uranium (greater than 93 percent) or with a few kilograms of plutonium-239 (also at least 93 percent). According to the International Atomic Energy Agency (IAEA), 25 kilograms (55 pounds) of U-235 or 9 kilograms (20 pounds) of Pu-239 represent the threshold amount of fissionable material required to build a nuclear weapon.

12. Roy L. Nersesian, *Energy for the 21st Century: A Comprehensive Guide to Conventional and Alternative Sources* (Armonk, NY: M. E. Sharpe, 2007), 276.

13. Although more abundant and potentially cheaper, Thorium is not directly fissionable. When it absorbs neutrons, it transforms largely into U-233, which is fissionable (thus Thorium is defined as *fertile*). Because of the high radioactivity of U-233, the need to chemically separate it from the irradiated thorium fuel and feed it to another reactor in a closed fuel cycle, fuel fabrication from thorium is very expensive. U-233 always contains traces of U-232, which in turn yield other isotopes that are strong gamma emitters with very short half-lives. Although this confers proliferation resistance to the fuel cycle, it results in increased costs. The concerns over weapon proliferation stem from the fact that U-233 can be separated on its own.

14. Treating the nuclear waste in closed-cycle power plants uses chemical processes that create *mixed oxides* of uranium and plutonium, indicated with the abbreviation MOX (Mixed Oxides, $UO_2 + PuO_2$).

15. In the fission of heavy elements, the combined mass of the resulting elements is slightly lower than the starting mass, and the difference is also an enormous amount of energy.

16. Of course, it makes neutrons that irradiate materials, as does fission.

17. There are 285 PWRs in twenty-five countries and 91 BWRs in ten countries.

18. *Heavy water* consists of two atoms of deuterium and one of oxygen (D_2O). Compared to the water with which we are familiar (H_2O), there are two atoms of a heavy isotope of hydrogen, called deuterium, in place of the hydrogen. Heavy water has the same chemical properties as normal water, but is denser (1.10 kg/liter at 25°C).

19. In the EPR, cooling and moderation are both obtained using pressurized water. It was designed and developed mainly by the French company Framatome (now Areva NP) and the German company Siemens AG. The principal aims of the project are to increase safety and at the same time achieve better economic competitiveness by improving the previously tested technologies (PWR) to achieve a potential 1.6 GW and by giving the EPR a useful life of sixty years.

20. Most of the spent fuel is low-level radioactive waste. The real issue for nuclear power waste disposal is the content of the irradiated fuel. There are three major components:

(1) The remaining uranium, with some different isotopic composition, which makes up almost all of the fuel mass and is not highly radioactive.
(2) The fission products, which are highly radioactive. The major components have half-lives of about 30 years and the waste is manageable after many half-lives, that is, a few hundred years.
(3) The transuranic elements, including plutonium and the minor actinides. Many of these have half-lives of thousands of years.

The U.S. plan is to dispose of the entire spent fuel package, whereas the French plan is to deal with the fission products and the minor actinides. In either case, the problem is the high-level waste.

21. International Atomic Energy Agency (IAEA), *Nuclear Power Reactors in the World* (Vienna: IAEA, 2006), 374; Uranium Information Centre, "Decommissioning Nuclear Facilities," Nuclear Issues Briefing Paper 13, June 2007.

22. For up-to-date information about dismantling power plants, see the website of the U.S. Nuclear Regulatory Commission at www.nrc.gov. For the rest of the world, go to the website of the World Nuclear Association at www.world-nuclear.org.

23. In the United States, initial licenses are for forty years, with most reactors receiving license extensions of at least twenty (maybe forty) years.

24. The *overnight cost* is the cost of building a plant as if it were completed "overnight," in other words if no time-related costs, such as interest and unforeseen costs, were incurred during construction.

25. Massachusetts Institute of Technology, *2007 Update of The Future of Nuclear Power: An Interdisciplinary MIT Study*, available at http://web.mit.edu/nuclearpower/pdf/nuclearpower-update2009.pdf, p. 6.

26. Massachusetts Institute of Technology, 2003. *The Future of Nuclear Power: An Interdisciplinary MIT Study*. Cambridge: The MIT Press, p. 38.

27. Cambridge Energy Research Associates (CERA), *Capital Costs Analysis: Power Market and Index Forecasts* (Cambridge, MA: CERA, 2009).

28. CERA, *Capital Costs Analysis*, 42. Costs were calculated assuming the real prices of the respective fuels for forty years. The price of coal

is $1.20 per million British thermal units; the equivalent unit price of gas is $5.40.

29. With a $50 carbon tax, the price per kilowatt-hour for coal would rise to 5.8 cents; gas would rise to 4.8 cents.

30. Vaclav Smil, *Energy at the Crossroads: Global Perspectives and Uncertainties* (Cambridge, MA: MIT Press, 2003), 85.

31. Building a combined-cycle gas turbine plant typically takes two or three years, a wind power installation one or two years, and a coal-fired power plant about four years. IEA, *World Energy Outlook 2006*, 372.

32. IAEA, *Nuclear Power Reactors in the World*.

33. In Japan, the average time is 60 months, and in China some plants were finished in 48 months.

34. The waste from power plants includes not only exhausted fuel but also resins, various radioactive consumable materials, and so forth, which are usually packed in concrete or compacted and stored. Their density is between 2,000 and 3,000 kg/m^3.

35. So far, the two European treatment centers have accumulated about 6,000 cubic meters of highly radioactive waste, which increases each year by 240 cubic meters, or about 1,600 tonnes. The density of these wastes is between 6,000 and 7,000 kg/m^3. They consist of fission products and zircalloy sheaths (zircalloy is widely used in the nuclear industry to coat the fuel rods).

36. Smil, *Energy at the Crossroads*, 312.

37. The total cost of decommissioning depends on the sequence and the timelines for the various phases of the program. The expense data shown here were taken from IEA, *World Energy Outlook 2006*, 374, and from Maurizio Cumo et al., *Nuclear Plants Decommissioning* (Rome: La Sapienza University, 2002), 379–81.

38. IAEA, *International Status and Prospects of Nuclear Power* (Vienna: IAEA, 2009).

39. World Nuclear Association, *The Global Nuclear Fuel Market: Supply and Demand, 2007–2030* (London: World Nuclear Association, 2009).

Chapter 6

1. In 2005, world gross production was 18,307 terawatt-hours (TWh) (about 18,235 TWh after taking out pumping requirements), to which hydroelectric energy contributed 2,994 TWh (about 2,922 TWh net of pumping). See International Energy Agency (IEA), *Electricity Information 2007* (Paris: IEA, 2007).

2. It is good to remember that many of these dams were built not for electrical production but to collect water for irrigation and to supply industrial and civilian consumption.

3. Patrick McCully, *Silenced Rivers: The Ecology and Politics of Large Dams* (London: Zed Books, 2001), 74–75. This book contains one of the best historical

reconstructions of the issues associated with the construction of large dams and the political, social, and economic problems they have brought about.

4. World Commission on Dams, *Dams and Developments* (London: World Commission on Dams, 2000). The commission completed its work in two years, as required by its constitution; see also www.dams.org.

5. International Energy Agency (IEA), *Electricity Information 2007* (Paris: IEA, 2007).

6. Dams are classified in three categories: *majors* (also called *superdams*), *larges*, and *smalls*. There are different criteria for identifying the categories of a dam, referring to height, fill capacity, flow volume, electrical capacity, and the size of the area it serves. Generally, a large dam is at least 50 feet (15 meters) high from base to crest. Superdams must meet at least one of the following four characteristics: a height of 500 feet (150 meters); a flow volume of at least 15 million cubic meters per hour of water; a reservoir capacity of 25 million cubic meters, or a generating capacity of 1 GW.

7. Located in Hubei Province in central China, the Three Gorges Dam is more than 7,500 feet (2,300 meters) long and 600 feet (185 meters) high, with a reservoir capable of holding more than 39 billion cubic meters of water. After a long planning process that began in 1992, construction began in 2003, to bring the dam to full operation in 2009 at an estimated cost of $28 billion. With an installed capacity of 18.2 GW, the dam should be able to generate almost 85 TWh per year.

8. *UDI World Electric Power Plants Database* (New York: Platts, 2006).

9. World Energy Council (WEC). *2007 Survey of Energy Resources*, pp. 271 et seq.

10. "China Recognizes Dangers Caused by Three Gorges Dam," *Wall Street Journal*, September 27, 2007.

11. With a final estimated cost of $32 billion, the dam system on the Tigris and Euphrates rivers would serve several purposes, including irrigation, water for civilian use, and the production of energy. When operational, its electrical generation capacity should be 7.4 GW using nineteen electrical power plants. Technical problems, safety issues, and a lack of adequate financial resources have caused the completion date to slip beyond the original objective of 2010.

Chapter 7

1. International Energy Agency (IEA), *World Energy Outlook 2006* (Paris: IEA, 2006).

2. Black carbon should not be confused with *carbon black*, which is produced from petroleum and is used in vehicle tires, as a pigment, and in photocopier toners.

3. Elisabeth Rosenthal, "Third-World Stove Soot Is Target in Climate Fight," *New York Times*, April 15, 2009 [interview with Dr. Veerabhadran

Ramanathan]. Available at http://www.nytimes.com/2009/04/16/science/earth/16degrees.html?_r=2.

4. IEA, *World Energy Outlook 2008* (Paris: IEA, 2008). Losses account for the remaining 20 percent.

5. In the United States, the Energy Independence and Security Act of 2007 set an annual production goal of 36 billion gallons of ethanol by 2022, 21 billion gallons of which was to come from cellulosic and other advanced biofuel production sources. In Europe, according to the new directive on the promotion of the use of energy from renewable sources, by 2020 each EU member state has to reach at least a 10 percent share of energy from renewable sources based on final consumption in the transportation sector. For the purposes of complying with the targets, the energy content of advanced biofuels is considered twice that of first-generation biofuels.

6. Bioethanol can be used to produce ethyl tributyl ether (ETBE), a chemical derivative of ethanol, which is used to increase the octane rating and oxygen content of gasoline. ETBE is gradually replacing methyl tributyl ether (MTBE), which has already been banned in much of the United States because of the contamination of groundwater. In the European Union in particular, bioethanol is used mainly to produce bio-ETBE, which can be mixed with gasoline. The current maximum standard is 22 percent.

7. *Transesterification* is the reaction of a triglyceride (fat or oil) with an alcohol to form esters and glycerol.

8. *Hydrolysis* is a process in which enzymes and other microorganisms are used to convert the cellulose and hemicellulose components of feedstock into sugars prior to their fermentation to produce ethanol.

9. *Biomass-to-liquids* is a thermo-chemical process in which pyrolysis/gasification technologies produce a synthesis gas ($CO + H_2$) from which a wide range of long-carbon-chain biofuels, such as synthetic diesel or aviation fuel, can be reformed.

10. Food and Agriculture Organization (FAO), *The State of Food and Agriculture* (Rome: FAO, 2008).

11. U.S. Department of Agriculture (USDA), Foreign Agriculture Service, *Grain: World Markets and Trade* (Washington, DC: USDA, 2009).

12. Based on data from the Energy Information Administration (U.S. Department of Energy) and the Renewable Fuel Association.

13. If we were to use sunflowers, the crop with the highest yield in Italy, this percentage would increase to 13 percent.

14. Following the chain from biomass to biofuels, there are the transformation processes to make biofuels from vegetable oil or sugars in the biorefineries. These processes yield secondary products in addition to the biofuels. Furthermore, complete conversion of the raw materials is not always assured.

15. Richard Doornbosch and Ronald Steenblik, *Biofuels: Is the Cure Worse than the Disease?* (Paris: OECD, 2007), 18–19.

16. Andreas Schäfer et al., *Transportation in a Climate-Constrained World* (Cambridge, MA: MIT Press, 2009).

17. Sugarcane uses the least energy to make biofuel because the woody part of the cane, called *bagasse*, is used to generate process energy, thus contributing to a positive energy balance for the entire sugarcane–bioethanol production chain.

18. *Fuel oxygenates* are chemicals containing oxygen that are added to fuels, especially gasoline, to make them burn more efficiently. Adding oxygenates to gasoline boosts its octane level and reduces the atmospheric pollution from automobile emissions.

19. For a broad view of this problem, see UN-Energy, *Sustainable Bioenergy: A Framework for Decision Makers* (New York: United Nations, 2007). This paper shows clearly how it is necessary to consider the entire life cycle of biofuels to determine their environmental impact accurately.

20. According to the International Water Management Institute, it takes on average about 820 liters of irrigation water to produce one liter of biofuel. However, the amount of irrigation water needed to produce biofuels varies widely depending on the area of the world. In Europe, where rainfed rapeseed is used, the amount of irrigation water required is 22 liters per liter of biodiesel. In the United States, where mainly rain-fed corn is used, the irrigation water requirement is 400 liters per liter of bioethanol. On the other hand, in China and India, countries with rapidly growing populations, which already have a shortage of water, each liter of ethanol requires up to 2,400 liters and 3,500 liters of irrigation water, respectively.

21. David Pimentel and Tadeusz W. Patzek, "Ethanol Production Using Corn, Switchgrass, and Wood; Biodiesel Production Using Soybean and Sunflower," *Natural Resources Research* 14, no. 1 (March 2005): 65–76.

22. For example, the Pimentel and Patzek study assumed a quantity of nitrogen fertilizers much higher than that used in other studies.

23. Alexander E. Farrell, Richard J. Plevin, Brian T. Turner, Andrew D. Jones, Michael O'Hare, and Daniel M. Kammen, "Ethanol Can Contribute to Energy and Environmental Goals," *Science* 311, no. 5760 (January 27, 2006): 506–8.

24. Doornbosch and Steenblik, *Biofuels*.

25. Tiffany A. Groode and John B. Heywood, *Ethanol: A Look Ahead*, LFEE 2007–002 RP (Cambridge: Massachusetts Institute of Technology, 2007).

26. This is known as the "food, feed, or fuel" competition.

27. For a broad overview of this problem, see C. Ford Runge and Benjamin Senauer, "How Biofuels Could Starve the Poor," *Foreign Affairs* (May/June 2007): 41–53; or UN-Energy, *Sustainable Bioenergy*.

28. American farmers would welcome an increase in the price of corn. Forecasts of an exploding growth in the demand for ethanol would lead them to swap out their crops. See, for example, USDA predictions of the average increase in net earnings for American corn producers in U.S. Department of

Agriculture, *Agricultural Baseline Projections to 2016*, OCE-2007-1 (Washington, DC: USDA, 2007), table 8.

29. According to some studies, if more than 10 percent of a crop is dedicated to biofuel, its price will depend on the international price of petroleum. See M. Kojima, D. Mitchell, and W. Ward, *Considering Trade Policies for Liquid Biofuels*, Renewable Energy Special Report 004/07 (Washington, DC: World Bank, 2007).

30. The bushel is the unit of measurement for cereals used in the United States. A bushel of corn weighs 56 pounds (25.4 kilograms) and yields about 2.5 gallons (11.4 liters) of bioethanol.

31. U.S. Bureau of Labor Statistics. CPI Detailed Report, March 2008. Malik Crawford ed., Washington DC.

32. Ralph Sims, Michael Taylor, Jack Saddler, and Warren Mabee, *From 1st- to 2nd-Generation Biofuel Technologies: An Overview of Current Industry and RD&D Activities* (Paris: OECD/IEA, 2008).

33. Scott Baier, Mark Clements, Charles Griffiths, and Jane Ihrig, "Biofuels Impact on Crop and Food Prices: Using an Interactive Spreadsheet," International Finance Discussion Paper No. 967 (Washington, DC: Federal Reserve System, 2009).

34. OECD and FAO, *OECD-FAO Agricultural Outlook, 2007–2016* (Paris: OECD, 2007).

35. George W. Huber and Bruce E. Dale, "Grassoline at the Pump," *Scientific American* 301, no. 1 (July 2009): 52–59. The story is about Cello, a company that promised to produce second-generation biofuels from woody-cellulosic biomass using a revolutionary and cheap technology, which did not exist. A federal court in Mobile, Alabama, ordered Cello Energy of Bay Minette, Alabama, to pay $10.4 million in punitive damages for fraudulently claiming it could produce cheap diesel-like fuel from hay, wood pulp, and other waste.

36. Schäfer, *Transportation in a Climate-Constrained World*.

37. IEA, *World Energy Outlook 2008*.

38. European Environment Agency (EEA), "Transport Biofuels: Exploring Links with the Energy and Agriculture Sectors," EEA Briefing 04 (Copenhagen: EEA, 2004).

39. John R. Benemann, *State of the Art of Microalgae Biofuels* (2009). Available at http://www.oilgae.com/blog/2008/09/dr-john-r-benemann-new-white-paper.html.

Chapter 8

1. In Germany, the wind world leader until 2007, total electricity production increased between 1990 and 2007 by about 90 TWh/year (16 percent), according to the International Energy Agency (IEA). About 45 percent of that increase (39 TWh/year) was supplied by wind power, which has achieved a share of more than 6 percent. During the same time,

photovoltaic solar power, which has also been heavily subsidized, covered only 4 percent of the increase, reaching a share of 0.5 percent.

2. 2006 data. IEA, *Energy Balances* (Paris: IEA, 2006).

3. Global Wind Energy Council (GWEC), *Global Wind 2008 Report* (Brussels: GWEC, 2008).

4. Offshore wind turbines may have higher capacity factor. Full-load hours for Danish offshore wind farms are in the range of 3,500 to 4,000 hours per year.

5. IEA Wind, *IEA Wind Energy Annual Report 2007* (Boulder, CO: PWT Communications, 2008).

6. GWEC, *Global Wind 2008 Report.*

7. In 2008, the U.S. wind industry installed 8.4 GW, an increase in generating capacity of 50 percent in a single calendar year. The industry has grown an average of 32 percent annually for the past five years, according to the GWEC.

8. In 2008, GWEC figures report that new installed capacity in China totaled 6.3 GW, a 91 percent increase over the 2007 market. The country's total capacity doubled for the fourth year in a row.

9. GWEC, *Global Wind 2008 Report.*

10. Wind turbines are also called *aerogenerators*. The wind energy collected by the vanes in mechanical form is transmitted through the hub. The hub is connected to a system with two shafts. The first is called the slow-speed shaft. It moves at the same angular speed as the rotor (a few dozen rpm, depending on the speed of the wind), transmitting the rotation to a series of gears connected directly to the other shaft, called the high-speed shaft. This shaft has the necessary angular velocity to operate the generator correctly (typically on the order of 1,500 rpm).

11. For example, an average wind speed of 5 m/s corresponds to an average power of 146 W per square meter, which is the velocity cubed times the correction coefficient.

12. European Wind Energy Association (EWEA), *Offshore Statistics 2009* (Brussels: EWEA, 2009).

13. "UK Anti-Wind Farm Groups Form Alliance against Development," *Platt's Renewable Energy Report*, June 29, 2009, 23.

14. The potential wind-power cost includes the cost of connecting to the existing transmission system, assuming that 10 percent of network capacity is available for new wind generation (not including new transmission lines to deliver power to distant markets). U.S. Department of Energy (DOE), *20 Percent Wind Energy by 2030: Increasing Wind Energy's Contribution to U.S. Electricity by 2030* (Washington, DC: DOE, 2008).

15. Energy Information Administration (EIA), *Assumptions to the Annual Energy Outlook 2009* (Washington, DC: EIA, 2009); IEA Wind, *IEA Wind Energy Annual Report 2007.*

16. DOE, *20 Percent Wind Energy by 2030.*

17. The Obama administration has made renewable power generation a key focus of its energy platform, with a national renewable power goal of 15 percent by 2020. The American Recovery and Reinvestment Act (ARRA) of February 2009 extends production tax credits for wind through 2012 (currently worth $21 per megawatt-hour for ten years) and grants wind developers the option of tacking on a 30 percent investment tax credit in lieu of the production credits. ARRA also supports financing of new wind projects and investment in new transmission grids to facilitate the expansion of renewable electricity generation. See Cambridge Energy Research Associates (CERA), *The U.S. Green Fiscal Stimulus* (Cambridge, MA: CERA, 2009).

18. To achieve the 20 percent target, wind is expected to contribute 12 percent of EU electricity by 2020. Commission of the European Communities, *Renewable Energy Road Map*, COM (2006) 848 (Brussels: European Commission, 2007).

19. The estimate was made by the Joint Coordinated System Plan, an organization involving major U.S. electrical system operators. See also Rebecca Smith, "New Grid for Renewable Energy Could Be Costly," *Wall Street Journal*, February 9, 2009.

20. In building wind farms, a minimum distance is needed from one wind turbine to the next, to avoid having one turbine block the wind to the other. The distance between two adjacent generators is a function of the rotor diameter. When wind farms take advantage of the prevailing wind direction, the distance between two generators typically is eight times the rotor diameter of the installations. When deployed at right angles to the prevailing wind direction, a distance of five times the rotor diameter is sufficient. This complication limits the density of wind farms in a given area.

Chapter 9

1. A great deal of information on the history of solar technology development can be found in: John Perlin, *From Space to Earth: The Story of Solar Electricity* (Ann Arbor, MI: AATEC, 1999).

2. The discovery of the properties of selenium was a classic case of serendipity. Selenium was used in a device to identify possible defects in cables when laying the first transatlantic telegraph cable in 1867. It was noticed that it only worked correctly in the absence of light.

3. The efficiency of these new selenium cells was not even 1 percent.

4. Maritime navigation lights were one of the first important applications of photovoltaic power, both in Japan (where Sharp installed exorbitantly expensive space-platform type cells), and in the United States, where the Coast Guard has imposed the use of navigation lights on oil platforms in the Gulf of Mexico since the end of the 1940s. These installations needed cumbersome and expensive batteries. Since the end of the 1960s, photovoltaic modules have solved many problems. The oil companies—Amoco,

Arco, Chevron, Exxon, Texaco, and Shell—bought them. At the end of the 1970s, Arco bought one of the module manufacturing companies, which later became Siemens Solar, then Shell Solar, and today SolarWorld.

5. The installation near Seville is part of a larger 300-MW project that will cost a little more than €1 billion and be finished in 2013. The builder is Solucar, of the Abengoa Group.

6. SolarBuzz: MarketBuzz 2009 at www.solarbuzz.com. The website displays a summary free of charge.

7. On the other hand, managing thermal gradients has been extensively studied to integrate photovoltaic power into buildings, for example, with solar chimneys or forced ventilation systems that use excess heat to warm or cool attics.

Chapter 10

1. For a more systematic discussion of geothermal energy, see Jefferson W. Tester et al., *Sustainable Energy* (Cambridge, MA: MIT Press, 2005), 453–510, and Mary H. Dickson and Mario Fanelli, "What Is Geothermal Energy?" (2004), http://iga.igg.cnr.it/geo/geoenergy.php.

2. This means that at 10,000 feet (3,000 meters) down, in the absence of geothermal anomalies, the subsoil has a temperature of about 212°F (100°C).

3. Production was declining in the fields at The Geysers and at Larderello. In both cases, reinjection wells were drilled, which allowed The Geysers to reactivate installations that were no longer productive and for Larderello to increase installed capacity.

4. The World Energy Council estimated that the worldwide potential is 35–75 GW.

5. The largest civilian heating installation in the world is at Reykjavík, Iceland. It can satisfy the needs of 160,000 people, almost all the inhabitants of the city.

6. International Energy Agency (IEA), *Geothermal Implementing Agreement, 2007 Annual Report* (Paris: IEA, 2007).

7. IEA, *Energy Balances* (Paris: IEA, 2006).

8. In essence, these heat pumps can be used to transfer energy from relatively cold bodies that can be cooled further to warmer bodies, the temperature of which can be increased using *geothermal probes*—wells a few dozen or a few hundred meters deep, into which a U-shaped tube is inserted carrying a fluid. This draws energy from the surrounding earth during its round-trip from the heat pump on the surface.

9. With suitable modifications, the heat pump can be used in the summer for cooling the same buildings, like a common air conditioner.

10. High-voltage power lines can connect national grids to power plants in remote areas, but heat distribution is somewhat limited.

11. IEA, *Geothermal Implementing Agreement, 2007 Annual Report*, 217.

12. At least with regard to electrical and thermal demand and limited to applications with temperatures compatible with the resources that can be exploited.

13. For an in-depth study of the long-term prospects of geothermal energy, see Tester et al., *Sustainable Energy*, 453–510.

14. Massachusetts Institute of Technology, *The Future of Geothermal Energy: Impact of Enhanced Geothermal Systems on the United States in the 21st Century* (Idaho Falls, ID: U.S. Department of Energy, Renewable Energy and Power Department, 2006).

15. Of special note is the very promising activity financed by the European Union at Soultz on the Franco-German border. Here, the temperature at 3 miles (5 kilometers) deep is 390°F (200°C).

16. The Obama administration took some steps in this direction with the March 2009 announcement by Energy Secretary Chu about the release of two Funding Opportunity Announcements for up to $84 million to support the development of EGS.

Chapter 11

1. The target of the European Platform is to get the cost of hydrogen under €2.50 per kilogram. One kilogram (2.2 pounds) of hydrogen is equal to 2.75 kilograms (6 pounds) of gasoline. See the European Commission's Fuel Cells and Hydrogen Joint Undertaking website at www.HFPeurope.org.

2. In a study published in *Journal of Power Sources* in 2002, the following costs for producing and transporting 900 kilograms of H_2 per day were hypothesized, using three options:

- Electrolysis *in loco* (the most expensive option, $6.50 per kilogram, but the only one potentially using renewable sources)
- Localized steam reforming with storage in gas bottles (the most economical option, $2.60 per kilogram)
- Centralized steam reforming plus liquefaction, transportation and reprocessing at the distribution site, with immediate costs but also opportunities for economies of scale

R. Mercuri, A. Bauen, and D. Hart, "Options for Refuelling Hydrogen Fuel Cell Vehicles in Italy," *Journal of Power Sources* 106 (2002): 353–63.

3. The typical distance used in these calculations was 80 kilometers (50 miles).

4. Essentially three factors drive the high cost of fuel cell vehicles:

- The fuel cell (about $17,500 more than a conventional car battery)
- The tank and the distribution system (several thousand dollars more than the traditional tank and distribution system)
- Scale

5. The Chevrolet Sequel from General Motors (GM), the first hydrogen-fueled car with a range of around 300 miles (500 km), has a tank holding 8 kilograms of hydrogen at high pressure (700 bars). The Chevrolet Equinox has a tank holding 4.2 kilograms and a range of 200 miles (320 km). The Opel Zafira HydroGen3 (mentioned on the website but not in commerce) has an optional tank holding 3.1 kilograms of hydrogen at 700 bars and a range of 170 miles (270 km), or a tank with 4.6 kilograms of liquid hydrogen and a range of 250 miles (400 km). At the Shanghai Car Show in Spring 2007, GM also presented the fifth-generation Chevrolet Volt E-Flex, half the size with the same performance. See www.gm.com and www.h2it.org.

6. Simple hydrides such as lithium hydride or beryllium hydride are caustic or toxic; those like titanium hydride are too stable. Some of these compounds, such as lithium-aluminum hydride, or better yet palladium hydride, are promising but not cheap.

7. The research into new structures includes metal organic frameworks, a new class of materials (Nathaniel L. Rosi, Juergen Eckert, Mohamed Eddaoudi, David T. Vodak, Jaheon Kim, Michael O'Keeffe, and Omar M. Yaghi, "Hydrogen Storage in Microporous Metal-Organic Frameworks," *Science* 300, no. 5622 [May 16, 2003]: 1127–29); fullerenes activated with lithium (Virginia Commonwealth University, July 2006); and more recently, carbon nanohorns (Centre National de la Recherche Scientifique, France, June 2007).

8. The implementation plan for the European hydrogen and fuel cells platform provides for spending about €2.2 billion for the research and demonstration program up to 2015 for the "vehicles and service stations" area, part of a much larger proposal concerning manufacturing technologies and using hydrogen for electrical energy. The total is estimated at more than €7 billion (2007–2015). The program should lay the groundwork for reaching the target by 2020 to produce and consume 2.2 million tons of hydrogen per year. Of this, 30 percent would be for the transportation sector (about 2 million automobiles sold each year), 45 percent for power generation, and 25 percent for portable micro fuel cells.

9. In the newly released budget, the U.S. Department of Energy cut $100 million from the Hydrogen Fuel Cell Program in fiscal year 2010 and transformed its name to Fuel Cell Technologies. Hydrogen, of course, is just the fuel of a fuel cell, a device that combines hydrogen and oxygen to produce water and electrical current. Still, the name change distances the Obama administration from the "hydrogen economy" goals of its predecessors.

10. Two-thirds of the *Hindenburg*'s passengers survived; there were thirty-nine fatalities. Most of the victims were killed by burning diesel fuel from the propulsion system. Meanwhile, the hydrogen rose rapidly upward, serving as a sort of parachute, which may have helped save the surviving passengers.

11. Other opportunities offered by a solar source are scission at ambient temperature in photo-electrochemical devices or using photosynthesizing catalytic converters. These are fascinating, but still in a research phase.

12. Hydroelectric power could be an exception. Even so, electrolysis is currently the most expensive option for producing hydrogen.

13. See the New York State Energy Research and Development Authority website, www.nyserda.org, for an overview. One thermochemical proposal is the S-I process (sulfuric acid–hydriodic acid) invented by General Atomic during the 1970s, using thermal energy at about 800–1,000°C from nuclear power.

14. With regard to the cost, it will be difficult for present nuclear technology to get below $14 per gigajoule, corresponding to about $1.75 per kilogram. The long-term prospects are for costs below $1.50 per kilogram, naturally with the advantage of being carbon-free. The problems with this technology are transporting fluids at high temperature and the development of low-cost catalyzers for the thermochemical process.

15. Leonardo Maugeri, *The Age of Oil* (Westport, CT: Praeger, 2006), 20–22, 43.

16. James J. Flink, *The Automobile Age* (Cambridge, MA: MIT Press, 2001), 10.

17. Anup Bandivadekar, Kristian Bodek, Lynette Cheah, Christopher Evans, Tiffany Groode, John Heywood, Emmanuel Kasseris, Matthew Kromer, and Malcolm Weiss, *On the Road in 2035: Reducing Transportation's Petroleum Consumption and GHG Emissions* (Cambridge: Massachusetts Institute of Technology, 2008).

18. Cambridge Energy Research Associates (CERA), *From the Pump to the Plug: What Is the Potential of Plug-in Hybrid Electric Vehicles?* (Cambridge, MA: CERA, 2008).

19. Henry Ford, *My Life and Work* (Garden City, NY: Doubleday, Page & Co., 1922).

20. R. J. Brood, "Factors Affecting U.S. Production Decisions: Why Are There No Volume Lithium-Ion Battery Manufacturers in the United States?" ATP Working Paper Series, Working Paper 05-01 (Gaithersburg, MD: Advanced Technology Program, NIST, 2005).

Chapter 12

1. E. Jochem, ed., "Energy End-Use Efficiency," in José Goldemberg, *World Energy Assessment: Energy and the Challenge of Sustainability* (New York: United Nations Development Program, 2000).

2. David Roland-Holst, *Energy Efficiency, Innovation, and Job Creation in California* (Berkeley, CA: Center for Energy, Resources, and Economic Sustainability [CERES], 2008).

3. American Council for an Energy-Efficient Economy, "The Positive Economics of Climate Change Policies: What the Historical Evidence Can Tell Us," Report E095, 2009. See www.aceee.org.

4. Commission of the European Communities, *Action Plan for Energy Efficiency: Realising the Potential*, COM (2006)545 (Brussels: Commission of the European Communities, 2006).

5. Hannah Choi Granade, Jon Creyts, Anton Derkach, Philip Farese, Scott Nyquist, and Ken Ostrowski, *Unlocking Energy Efficiency in the U.S. Economy* (New York: McKinsey & Co., 2009).

6. Granade et al., McKinsey Global Energy and Materials. *Unlocking Energy Efficiency*, xii.

7. International Energy Agency (IEA), *World Energy Outlook 2007* (Paris: IEA, 2008).

8. IEA, *Energy Technology Perspectives* (Paris: IEA, 2008).

9. IEA, *World Energy Outlook 2007*.

10. International Association of Public Transport (UITP), "Public Transport, the Green and Smart Solution: A New Frontier: Double Market Shares by 2025" (2008), available at http://www.uitp.org/advocacy/pdf/new_pt_strategy.pdf.

11. Amory Lovins published his first, seminal work about energy efficiency in 1977. See Amory Lovins, *Soft Energy Paths: Toward a Durable Peace* (Cambridge, MA: Ballinger Publishing, 1977). Eventually, he defended his views in many other works, among them: Ernst von Weizacker, Amory B. Lovins, and L. Hunter Lovins, *The Factor Four: Doubling Wealth, Halving Resource Use* (London: Earthscan, 1997); and Amory Lovins et al., *Winning the Oil Endgame: Innovations, Profits, Jobs and Security* (Snowmass, CO: Rocky Mountain Institute, 2004).

12. Mark Jaccard, *Sustainable Fossil Fuels* (Cambridge: Cambridge University Press, 2005), 95.

13. Summary of Comments by The Honorable Branko Terzic, 27th USAEE/International Association for Energy Economics, North American Conference September 18, 2007, Houston.

14. Terzic, ibid.

15. Peter R. Odell, *Why Carbon Fuels Will Dominate the 21st Century's Global Energy Economy* (Brentwood, England: Multi-Science, 2004), 4.

16. Odell, *Why Carbon Fuels Will Dominate*, 9.

Conclusions

1. See: Shoibal Chakravarty, Ananth Chikkatur, Helen de Coninck, Stephen Pacala, Robert Socolow, Massimo Tavoni, *Sharing Global CO_2 Emissions Reductions Among One Billion High Emitters*, National Proceedings of the Academy of Sciences of the United States of America *PNAS* 106, no. 29 (July 21, 2009): 11884–11888. See also at: http://www.pnas.org/content/106/29/11884.full.pdf+html?sid=10c4058a-8e28-4020-adec-d1039726e8a3.

Glossary of Units of Measurement

Barrel (b or bbl)

Standard unit of measurement corresponding to a volume of roughly 42 U.S. gallons, 35 Imperial gallons (Canada, UK), or 158 liters.

Barrel of Oil Equivalent (BOE)

Conventional unit of measurement of energy used to express different energy sources (typically fuels such as gas and oil) in a common unit of measurement, by taking into account their specific calorific value. It is assumed that a barrel of oil corresponds to 5.8 million BTUs.

By comparison, 1 BOE equals approximately 145 cubic meters of natural gas with a calorific power of 40,000 BTU/m^3.

Barrel per Day (b/d or bbl/d)

Unit of measurement of the production capacity of an oil well or oil field or of the processing capacity of a plant. A *day* can be a *solar day* or a *working operational day*. Using solar days, 1 b/d equates to 50 tons per year. The most-used multiples of b/d are thousands of barrels a day (kb/d) and millions of barrels a day (mb/d).

British Thermal Unit (BTU)

Unit of measurement of energy corresponding to the quantity of heat needed at standard atmospheric pressure to raise the temperature of one pound of water by one degree Fahrenheit, from 59.5°F to 60.5°F. A commonly used BTU multiple is millions of BTUs (MBTU).

Cubic Meter (m^3)

Unit of measurement of volume equal to a cube with one-meter sides. Cubic meters are commonly used for the measurement of natural gas under standard atmospheric pressure conditions (1 bar) and at the temperature of 59°F. The energy content of 1 m^3 of gas may vary from field to field depending on its precise mixture of constituent gases. For example, if 1 m^3 of gas from Algeria contains 42,000 kJ and 1 m^3 of gas from Russia contains 37,000 kJ, this means in energy terms that 1 m^3 of gas from Algeria is the equivalent of about 1.13 m^3 of gas from Russia.

Currencies

Unless otherwise indicated, the following symbols are used for currencies (ISO symbol):

$ — U.S. dollar (USD)
€ — Euro (EUR)
£ — British pound sterling (GBP)

Joule (J)

Unit of measurement of energy for the work, and quantity of heat equivalent to the work, produced by the power of 1 Newton when its point of application is moved by one meter in the direction of the power. The joule is a very small unit of energy, and hence its multiples are very often used: the kilojoule (kJ; thousands of joules), the megajoule (MJ; millions of joules), the gigajoule (GJ; billions of joules), and the terajoule (TJ; trillions of joules).

Tonne of Coal Equivalent (TCE)

Conventional unit of measurement based on the assumption that 1 kilogram of coal can yield 7,000 kilocalories (kcal)—hence, 1 TEC = 7×10^6 kcal. By comparison:

- 1 TCE = 0.7 TOE
- 1,000 m^3 of natural gas = ~1.2 TCE

The most-used multiples of TCE are the kilotonne of coal equivalent (kTCE) and the megatonne of coal equivalent (MTCE).

Tonne of Oil Equivalent (TOE)

Conventional unit of measurement used to express all energy sources with a common unit of measurement, taking into account their specific calorific value. It is assumed that a kilogram of oil corresponds to 10,000 kilocalories (kcal), so 1 TOE = 10^7 kcal.

- 1 TOE = ~1.4 TCE
- 1,000 m^3 of natural gas = ~0.83 TOE

The most-used multiples of the TOE are thousands of TOE (kTOE) and millions of TOE (MTOE).

Watt (W)

Unit of measurement of mechanical power, electrical power, work, and heat calculated in time units (1 W = 1 joule per second). The most-used multiples of the watt are the kilowatt (kW; thousands of watts), megawatt (MW; millions of watts), gigawatt (GW; billions of watts), and terawatt (TW; trillions of watts).

Watt-Hour (Wh)

Unit of measurement of energy, work, and heat equivalent to the energy produced by a 1-watt machine in 1 hour's time. The Wh is actually a very small unit of energy, hence its multiples are very often used: kilowatt-hours (kWh), megawatt-hours (MWh), gigawatt-hours (GWh), and terawatt-hours (TWh).

Index

About the Author and Translator

LEONARDO MAUGERI is Senior Executive Vice President of Strategies and Development at Eni SpA, one of the world's leading international oil and gas companies and Italy's largest industrial company. He directs Eni's research and technology policies and renewables programs, and he serves on the boards of directors of Polimeri Europa SpA, Italgas, and Fondazione Mattei. Internationally recognized as a preeminent expert on oil, gas, and energy, Dr. Maugeri is a Visiting Scholar of the Massachusetts Institute of Technology (MIT) and a member of MIT's External Energy Advisory Board. He is the author of four books on energy, including *The Age of Oil: The Mythology, History, and Future of the World's Most Controversial Resource"* (Praeger, 2006), which was named a CHOICE 2007 Outstanding Academic Title, won the 2007 Premio Roma, and has been translated into 11 languages. His articles on oil, gas, and energy issues have appeared in mainstream and professional periodicals, including *Foreign Affairs, Science, Scientific American, Forbes, The Oil & Gas Journal*, and *The Wall Street Journal*. Dr. Maugeri contributed the lead story of the 2006 energy issue of *Newsweek*.

JONATHAN T. HINE JR. grew up in Rome, Italy and translated his first book in 1961. A graduate of the U.S. Naval Academy (B.S.), the University of Oklahoma (MPA), and the University of Virginia (Ph.D.), he is a Life Member of the American Translators Association (ATA) and was the founding Secretary-Treasurer of the American Translation and Interpreting Studies Association. In addition to translating full-time, Dr. Hine conducts business and organization workshops throughout the United States and writes self-help books and articles. He is a frequent presenter at ATA conferences and an ATA mentor and certification grader. He lives in Charlottesville, Virginia with his family and loves riding his bicycle on the open road. Contact: hine@scriptorservices.com.

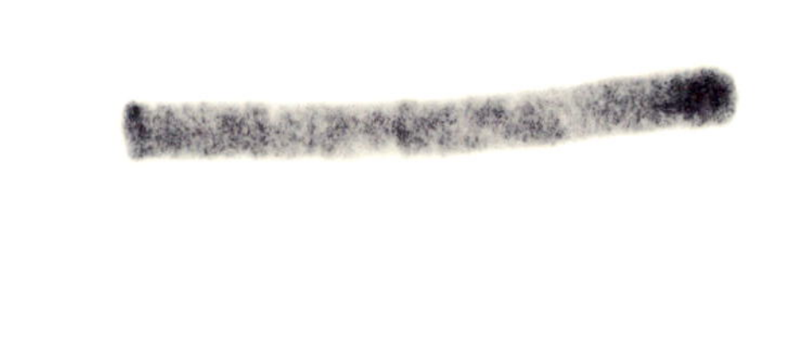

RECEIVED
MAY 14 2010